Working Fluid Selection for Organic Rankine Cycle and Other Related Cycles

Working Fluid Selection for Organic Rankine Cycle and Other Related Cycles

Special Issue Editor

Attila R. Imre

MDPI • Basel • Beijing • Wuhan • Barcelona • Belgrade • Manchester • Tokyo • Cluj • Tianjin

Special Issue Editor
Attila R. Imre
Budapest University of
Technology and Economics
Hungary

Editorial Office
MDPI
St. Alban-Anlage 66
4052 Basel, Switzerland

This is a reprint of articles from the Special Issue published online in the open access journal *Energies* (ISSN 1996-1073) (available at: https://www.mdpi.com/journal/energies/special_issues/working_fluid_selection_for_ORC).

For citation purposes, cite each article independently as indicated on the article page online and as indicated below:

LastName, A.A.; LastName, B.B.; LastName, C.C. Article Title. *Journal Name* **Year**, *Article Number*, Page Range.

ISBN 978-3-03936-074-1 (Hbk)
ISBN 978-3-03936-075-8 (PDF)

Contents

About the Special Issue Editor

Attila R. Imre received his M.Sc. in Physics from Eötvös University, Budapest in 1990; four years later, he obtained a Ph.D. from the same institution. He has spent several years in various U.S. and German universities, like the University of Tennesee, Johannes Gutenberg Universität Mainz, and Universität zu Köln. InHungary, he held various positions in the Atomic Energy Research Institute. Later, this institute became part of the newly formed Centre for Energy Research, where A.R. Imre is a Scientific Advisor and the Head of the Scientific Council of the research centre. After obtaining a Doctor of Science title from the Hungarian Academy of Science in 2015, he was appointed a Full Professor of Energy Engineering with the Budapest University of Technology and Economics. His present research topics cover various fields of energy production and storage, including the use of low-temperature heat sources (like geothermal, solar, and industrial waste heat) by ORC and similar methods and the application of power-to-gas technologies in energy storage. He has authored around 100 papers in various international journals.

Preface to "Working Fluid Selection for Organic Rankine Cycle and Other Related Cycles"

To create a more environmental-friendly and sustainable society, previously improperly used "low-quality" energy sources can be used. In the last few decades, power generation from low-temperature heat sources (below 300 °C), like thermal, solar, geothermal, biomass, or waste heat, has become increasingly significant. The traditional Rankine cycle using water as a working fluid cannot be used with sufficient efficiency at low temperatures. Therefore, finding novel working fluids for organic Rankine cycles or similar but less frequently used thermodynamic cycles (like trilateral flash cycles) has become a priority.

Traditionally, the working fluid for a given ORC process is selected using a trial-and-error procedure through experience from chemically similar materials. This method, however, risks excluding novel and previously unused working fluids that could be more suitable for the given heat source than any of the traditional fluids. In this Special Issue, more sophisticated methods are presented using optimization models, thermodynamic analyses, equation-of-state parameters, and molecular properties. Expanders and working fluids should be selected together; therefore, some discussion related to special ORC-expanders is also included.

We aim to present a reliable source for researchers, innovators, and developers working on ORC-related fields to help them to find the proper working fluid/expander pairs for any given heat source.

Attila R. Imre
Special Issue Editor

energies

MDPI

Article

Thermodynamic Selection of the Optimal Working Fluid for Organic Rankine Cycles

Attila R. Imre [1,2,*], Réka Kustán [1] and Axel Groniewsky [1]

1 Department of Energy Engineering, Budapest University of Technology and Economics, Muegyetem rkp. 3, H-1111 Budapest, Hungary; kustan@energia.bme.hu (R.K.); groniewsky@energia.bme.hu (A.G.)

2 Centre for Energy Research, Department of Thermohydraulics, Hungarian Academy of Science, P.O. Box 49, H-1525 Budapest, Hungary

* Correspondence: imreattila@energia.bme.hu or imre.attila@energia.mta.hu

Received: 6 May 2019; Accepted: 24 May 2019; Published: 27 May 2019

Abstract: A novel method proposed to choose the optimal working fluid—solely from the point of view of expansion route—for a given heat source and heat sink (characterized by a maximum and minimum temperature). The basis of this method is the novel classification of working fluids using the sequences of their characteristic points on temperature-entropy space. The most suitable existing working fluid can be selected, where an ideal adiabatic (isentropic) expansion step between a given upper and lower temperature is possible in a way, that the initial and final states are both saturated vapour states and the ideal (isentropic) expansion line runs in the superheated (dry) vapour region all along the expansion. Problems related to the presence of droplets or superheated dry steam in the final expansion state can be avoided or minimized by using the working fluid chosen with this method. Results obtained with real materials are compared with those gained with model (van der Waals) fluids; based on the results obtained with model fluids, erroneous experimental data-sets can be pinpointed. Since most of the known working fluids have optimal expansion routes at low temperatures, presently the method is most suitable to choose working fluids for cryogenic cycles, applied for example for heat recovery during LNG-regasification. Some of the materials, however, can be applied in ranges located at relatively higher temperatures, therefore the method can also be applied in some limited manner for the utilization of other low temperature heat sources (like geothermal or waste heat) as well.

Keywords: adiabatic expansion; isentropic expansion; T-s diagram; working fluid classification; optimization

1. Introduction

For converting the heat of low-temperature heat sources into electricity, the Organic Rankine Cycle (ORC) is often applied [1,2]. There are several known—and probably even more unknown—low boiling-point working fluids that can be used to convert heat into work. The selection of the ideal working fluid for a given heat reservoir requires a multi-dimensional optimization method; selection criteria include thermodynamic and physical properties, chemical stability, chemical compatibility, environmental impacts, safety, production costs, availability of the fluid and several other parameters [1,3,4]. Some of these parameters are more important than the others and even the rank of importance can be different in various locations or time, making the evaluation process even more difficult. Based on the evaluation of those criteria, one might obtain various indicators for each of these criteria, reflecting the performance of a given working fluid and the corresponding layout. Then these indicators can be combined [5]–with proper weight—to a comprehensive evaluation indicator to rank the working fluids and to find the best of them [6]. For this kind of evaluation, all sub-evaluation steps focusing only on a single aspect should be as accurate as possible.

For this reason, we are focusing on one segment of this optimization method, related to the selection of the proper working fluid to avoid wet or superheated vapour states at the end of the expansion process;

in this case, one might use the simplest ORC layout, using only four units, including a pump, a heat exchanger, an expander and a second heat exchanger. Starting the expansion from a saturated vapour state and using wet working fluid, the expansion is terminated in a mixed-phase state, in the so-called wet vapour region. Here, the vapour contains dispersed liquid (droplets), which can cause loss of efficiency as well as damage of expander due to droplet erosion. To avoid erosion problems, one can use a droplet separator device to remove the liquid part. Also, it is possible to apply superheating (i.e., increase of the fluid temperature from the saturated vapour state to a superheated dry vapour state); in this case, an additional heat exchanger section (superheater) is needed. Using dry fluid and starting the expansion from a saturated vapour state, the expansion has to be terminated in a low-temperature superheated dry vapour state. Here, some residual heat has to be removed from the working fluid before condensation. This can cause some problems, including loss of efficiency or high cooling load of the condenser. To avoid these problems, one might choose the inclusion of an extra heat exchanger, called recuperator, which cools the superheated vapour before reaching the condenser and preheats the liquid (i.e., recovers some of the lost heat) before reaching the evaporator. The best solution would be the application of an isentropic working fluid, where ideal adiabatic expansions could run along (or nearby) the saturated vapour states during the whole expansion process. In this case a simple set-up of two heat exchangers, a pump and an expander could be sufficient to convert the heat to work by ORC. Unfortunately, such an ideal isentropic fluid does not exist; all real fluids classified as isentropic one have a reverse S-shaped saturated vapour branch on *T-s* diagram, not a straight line with infinite slope, assumed by isentropicity. This is also true for fluids described by various molecular or equation of state based models [7–13].

Using our method, for a given heat source—heat sink pair one can select a real, one-component working fluid from a database [14], with an ideal adiabatic (isentropic) expansion process starting from a saturated vapour state and terminating also in a saturated vapour state (or at least in the vicinity of this state), utilizing the "belly" of the reverse S-shaped saturated vapour branches of various materials. In this way, one can use the simplest ORC layout, consisting of only a pump, two heat exchangers (evaporator, condenser) and an expander, avoiding the use of superheater or droplet separator and a recuperative heat exchanger (recuperator). Since the fluid is always in dry condition during expansion, droplet erosion of the expander can also be avoided. Concerning the description of different set-ups used with various working fluids, more details can be seen for example in [1,8].

Presently, we have a database of around 30 pure fluids with *T-s* data taken from the NIST Chemistry WebBook [15] and from RefProp 9.1 [16]. Only fluids with properly accurate *T-s* data are included into the database; they are the ones where *T-s* data are calculated by high-accuracy reference equations, instead of any other general equations of states. Most of these fluids were termed formerly as dry and a very few of them as isentropic, while in the novel classification they are in various isentropic sub-classes, namely in ANCMZ, ACNMZ, ANZCM and ANCZM.

From the present working fluid set, one can choose the thermodynamically most suitable material for a given ORC-cycle. Due to the temperature range covered by the *T-s* data, these fluids should be proper candidates mostly for cryogenic cycles, but after proper expansion, the database can be used for other temperature ranges (like geothermal, solar or waste heat applications) as well.

2. Maps of Potential Expansion Routes

According to the novel classification scheme, working fluids are classified by the entropy-sequences of their so-called primary and secondary characteristic points on the *T-s* diagram:

- Points A and Z are the initial and final points of the *T-s* saturation curve. Physically it is related to the triple point (or freezing point)
- Point C is the critical point, located on the top of the curve, separating the liquid branch (A-C) and the vapour branch (C-Z)
- Points M and N are local or global entropy extrema, located on the vapour branch.

Points A, Z and C exist for all working fluids (these are the primary points), while M exists only for dry and isentropic ones and N exists only for isentropic ones (therefore these are the secondary points). Theoretically, eight different classes can exist, one of them is wet (ACZ), two of them are dry (ACZM and AZCM) and five of them are "real" isentropics, with reverse S-shaped vapour branch (ANZCM, ANCZM, ANCMZ, ACNMZ and ACNZM). Concerning real materials, examples can be found only for six of these classes, neither ACZM, nor ACNZM have been found up to now. Some of these real isentropic fluids (characterized by five-letter sequences) were considered formerly as dry ones, being the point N well below ambient temperatures. Using this classification, one can easily correlate simple material properties like molar volume [12] or molar isochoric heat capacity [14] with the classification, i.e., separate wet and dry working fluids. Also, it can be used easily to study the behaviour of working fluids during the expansion process of ORC and other similar cycles [17].

The above mentioned reverse S-shape allows us to define isentropic expansion routes. Two different *T-s* diagrams are shown in Figure 1, for 2-methylbutane (a) and benzene (b). It can be seen, that although none of them is an isentropic working fluid according to the traditional wet-dry-isentropic classification, one can still define isentropic expansion routes starting and terminating on the saturated vapour branch; these routes are marked as 1→2 and 3→4. For both materials, all potential isentropic expansion routes located in a "bay", separated by a limiting expansion line (grey dashed one) from the rest of the diagram. For ANCMZ-type working fluids, the limiting line is defined by the $M \rightarrow M_s^d$ expansion route; this is also true for ACNMZ-types. For ANCZM types, the limiting line is defined by the $Z_s^u \rightarrow Z$ expansion route; the same is true for ANZCM types.

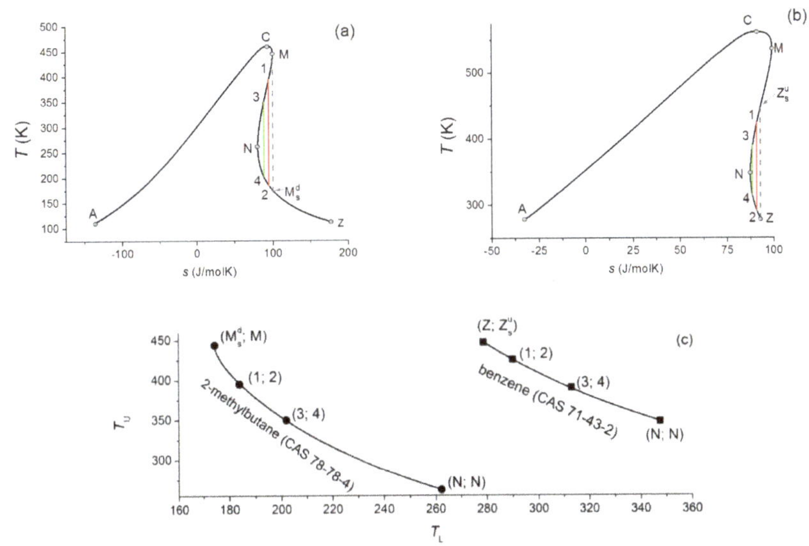

Figure 1. Potential isentropic expansion routes (1→2 and 3→4) for working fluids classified as ANCMZ (**a**; 2-methylbutane) and as ANCZM (b; benzene). Limiting expansion curves marked by dashed grey lines from $M \rightarrow M_s^d$ (**a**) and $Z_s^u \rightarrow Z$ (**b**). Figure (**c**) shows the way of constructing a T_L-T_U diagram representing expansion routes for each material. One material is represented by one curve and one point of the given curve illustrates one expansion process with a given upper (starting) and lower (finishing) temperature. Numbers and letters on (**a**,**c**) and (**b**,**c**) figures are corresponding ones.

Here, Z_s^u and M_s^d are the so-called tertiary characteristic points, marking the intersection on the saturation curve projecting the primary and secondary ones. Subscript s refers to the route of the projection (to the s-axis), while superscript u and d refers to the relative location compared to the original point (u: up; d: down).

Each expansion process can be represented by an upper and a lower temperatures (T_U and T_L), where expansion starts and stops. In this way, one can construct a T_L-T_U diagram (shown in Figure 1c) illustrating expansion routes for each material. One material is represented by one curve and one point of the given curve represents one expansion process with a given upper (starting) and lower (finishing) temperature. Numbers and letters on a-c (2-methylbutane) and b-c (benzene) figures are corresponding ones.

Concerning the *T-s* diagrams used here, the specific entropy values are given for moles, instead of mass (i.e., in J/molK, instead of J/kgK). Although it is not usual in engineering applications, it has been shown recently, that for the comparison of various working fluids, the mole-based quantities are more suitable than the mass-based ones [14].

Using 28 materials with reverse S-shaped vapour branch and with *T-s* diagrams available in the NIST databases (NIST Chemistry WebBook [15] and from RefProp 9.1 [16]), one can construct a map for all potential routes; this map is shown in Figure 2. The lower end for all lines are on one curve, defined by the N-points; for the shortest possible expansion route $T_U = T_N + \delta$ and $T_L = T_N + \delta$, where T_N refers to the temperature corresponding to point N and δ is an infinitesimally small temperature difference with $\delta \to 0$. For this reason, the lower limit should be located on a straight line with slope = 1, where $T_U = T_N = T_L$. Marking this lower limit, one can avoid unrealistic $T_L > T_U$ cases.

Materials are identified by their chemical identifier (CAS number). When lines are congested, labels are following the order of lines in zig-zag. Materials corresponding to various classes are plotted in various colors; black for ANCMZ, red for ACNMZ, green for ANZCM and blue for ANCZM; as it was already mentioned, materials with ACNZM sequences have not been found up to now. For additional information, Carnot-efficiencies for a given upper and lower temperature pair are given by color-coded lines. The list of materials with their CAS-number, name, refrigerant code (if any) and classification can be found in Table 1.

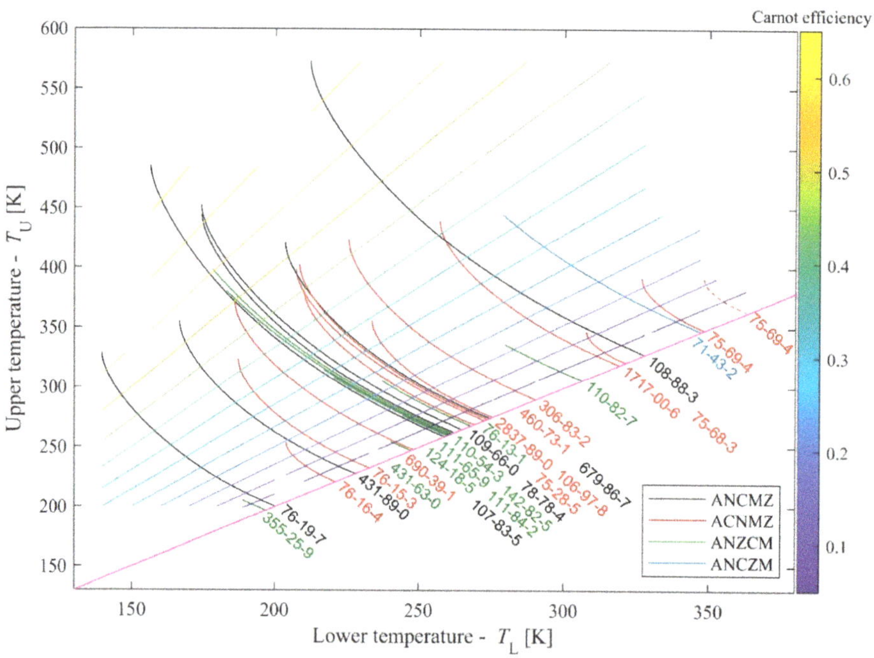

Figure 2. Potential isentropic expansion routes on T_L-T_U diagram for 28 working fluids, identified by their CAS-number. Carnot-efficiencies for a given upper and lower temperature pair are given by color-coded lines.

Since *T-s* data are available in various sources, NIST-based data (NIST Chemistry Webbook [15] and RefProp 9.1 [16]) has been cross-checked. Visible difference was found only in one occasion (CAS 75-69-4, trichloromonofluoromethane, also called R11 or Freon 11); the alternative curve (data are taken from FluidProp [18], using TPSI model, based on the work of Reynolds [19])) is also plotted by the dashed red line.

Carnot-efficiencies are also shown in Figure 2. These values are characteristic for the given heat source—heat sink pairs. Being the corresponding T_U and T_L values are external constraints, Carnot-efficiencies cannot be improved. They are shown here (and later in similar figures) to give some indication about the theoretical efficiency maximum as a benchmark.

Table 1. List of materials used in the study. The list follows the order of Figure 2 starting at decafluorobutane, marked as 355-25-9 to trichlorofluoromethane (also referred as R11), marked as 75-69-4.

CAS-Number	Name	R-Number	Class
355-25-9	decafluorobutane	-	ANZCM
76-19-7	perfluoropropane	R218	ANCMZ
76-16-4	freon 116	R116	ACNMZ
431-89-0	propane, 1,1,1,2,3,3,3-heptafluoro-	R227ea	ANCMZ
76-15-3	freon 115	R115	ACNMZ
431-63-0	1,1,1,2,3,3-hexafluoropropane	R236ea	ANZCM
690-39-1	1,1,1,3,3,3-hexafluoropropane	R236fa	ACNMZ
107-83-5	2-methylpentane (isohexane)	-	ANCMZ
124-18-5	decane	-	ANZCM
111-84-2	nonane	-	ANZCM
111-65-9	octane	-	ANZCM
142-82-5	heptane	-	ANZCM
110-54-3	hexane	-	ANZCM
78-78-4	2-methylbutane (isopentane)	R601a	ANCMZ
109-66-0	pentane	R601	ANCMZ
76-13-1	freon 113	R113	ANZCM
75-28-5	2-methylpropane (isobutane)	R600a	ACNMZ
106-97-8	butane	R600	ACNMZ
2837-89-0	freon 124	R124	ACNMZ
679-86-7	1,1,2,2,3-pentafluoropropane	R245ca	ANCMZ
460-73-1	1,1,1,3,3-pentafluoropropane	R245fa	ACNMZ
306-83-2	freon 123	R123	ACNMZ
110-82-7	cyclohexane	-	ANZCM
1717-00-6	1,1-dichloro-1-fluoroethane	R141b	ACNMZ
75-68-3	ethane, 1-chloro-1,1-difluoro-	R142b	ACNMZ
108-88-3	toluene	-	ANCMZ
71-43-2	benzene	-	ANCZM
75-69-4	trichlorofluoromethane	R11	ACNMZ

In the next sections, comparison of T_L-T_U diagram of real materials (where "real" refers to data taken from well-established reference equations) with T_L-T_U diagram of model fluids (van der Waals fluid) is given; this way one might be able to single out erroneous data. Then, the method to find the working fluid which can be used with the simplest layout for a given heat source—heat sink pair by using the T_L-T_U diagram of real fluids is presented.

3. Comparison of Reduced T_L-T_U Diagrams

The behaviour of fluids can be described with various equations of states (EoS); usually simple, easily solvable equations are giving less accurate descriptions than more complex, sometimes even non-analytical ones. Luckily, the results given by some of these simple EoS are qualitatively acceptable, therefore they can be applied for comparison purposes. Concerning *T-s* diagrams, two model fluids has been studied earlier, using van der Waals [8] and Redlich-Kwong [10] EoS. One of the advantages of using simple EoS is to reduce the possibility to obtain mathematical artefacts caused by some more accurate, but mathematically more complex EoS (see for example the misplaced second critical point given by the very accurate water EoS, the IAPWS [20]). Therefore, the van der Waals EoS with temperature independent molecular degree of freedom has been used to obtain the "theoretical" T_L-T_U diagram. The result can be seen in Figure 3. Model materials corresponding to various classes are plotted in various colors, also used for Figure 2. Concerning the qualitative agreement mentioned above, it is interesting to see, that the class missing in case of real materials (ACNZM) is also missing here.

One can realize, that the curves are located within a triangle. The lower boundary is defined by the coordinate of the N points, similarly to Figure 2. Concerning the upper boundary (magenta line), these are the T_L and T_U values for the "longest" expansion lines for classes ACNMZ (red) and ANCMZ (black), namely the points defined by temperatures corresponding to M_s^d (as T_L) and M (as T_U). The exact shape of this curve is not clarified, but it seems to be close to linear. Concerning the third side of the triangle, it is an artificial limit (therefore it is not marked by a line), caused by the fact that all *T-s* diagrams of van der Waals fluids have been calculated in the same reduced temperature range (from $T_r = 0.3$ to $T_r = 1$), where the reduced temperature is the actual temperature divided by the critical one. Although this limit seems to be artificial, it reflects correctly, that the "liquid range" of materials located between the triple point and the critical point is usually in the 0.4–1 reduced temperature range for most of the materials [21]. The temperature of these "lowermost points of the liquid state" are identical with the temperatures of point Z (and also with point A), therefore the "longest" expansion lines for classes ANCZM (blue) and ANZCM (green), are terminated on a line formed by the upper and lower expansion end-points, defined by temperatures corresponding to Z (as T_L) and Z_s^u (as T_U).

For comparison, the reduced T_L-T_U diagram for the real materials (shown in Figure 2) can be seen in Figure 4. The real curves are also located in a triangular region. The lower border is as sharp as before, because it is the slope = 1 line, defined by the identical T_U and T_L temperatures of points N. One can also see an upper boundary defined by temperatures corresponding to M_s^d (as T_L) and M (as T_U). While for the model fluid, this upper boundary for ANCMZ (black) and ACNMZ (red) classes is quite sharp, one can see that it is less defined here. Concerning the third side of the region, it can be noted that the expansion line for most of the materials in the ANZCM class (green) terminate in the reduced T_L = 0.35–0.4 region (as it is "expected" from the thumb-rule between melting and critical temperatures, see for example [21]), but the remaining part of the ANZCM-class materials as well as the only ANCZM-one terminate at higher reduced T_U temperatures, showing much narrower liquid region between freezing and critical point.

Concerning trichlorofluoromethane (CAS 75-69-4), which has a different representation using NIST-based data or Stanford (TPSI)-based data, it seems that both curves are too short (they terminate much below the upper boundary), but the TPSI-data seems to be better. For other materials, no visible differences have been found. From the data obtained by van der Waals EoS, one can see the existence of an upper envelope of the curves. Therefore, one can expect that for real materials, the curves will also be confined between an upper and a lower borderline. It means that for example shorter expansion lines of ACNMZ class fluids should be located on higher $T_{L,r}$ temperatures, than longer expansion lines. Hence, the short solid line representing trichlorofluoromethane (R11, 75-69-4) is probably misplaced and it should be located above the expansion line of freon-116 (76-16-4) and not below it. Consequently, the dashed red lines based on TPSI-data seems to be better than the one, based on NIST data.

On one hand, by using reduced temperature data, the structure of T_L-T_U diagram are much clearer now; also it can help to single out erroneous data. On the other hand, for the selection of the optimal working fluid for a given heat source—heat sink, the use of the original T_L-T_U diagram is necessary, as it will be shown in the next section.

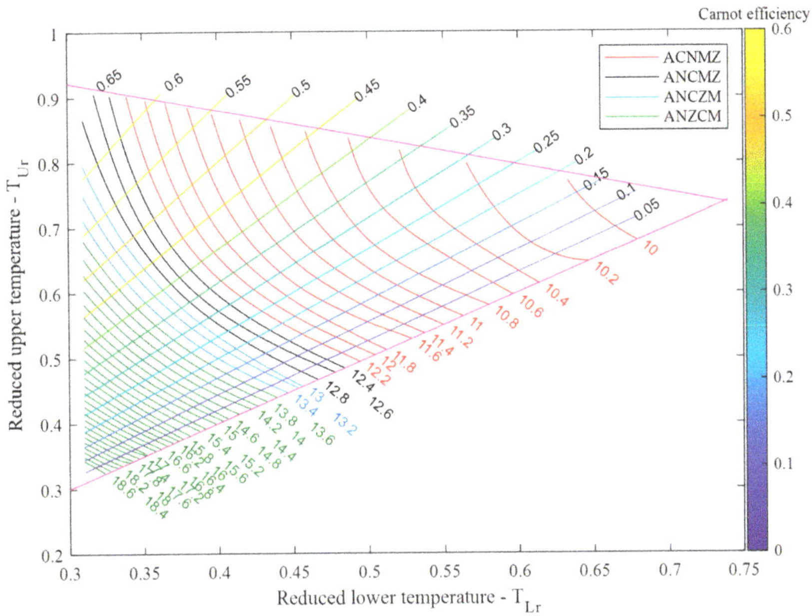

Figure 3. Potential isentropic expansion routes on T_L-T_U diagram for van der Waals model fluids. Molecular degree of freedom are marked for each curves at the bottom. Carnot-efficiencies for a given upper and lower temperature pair are given by color-coded lines with numerical values on the top.

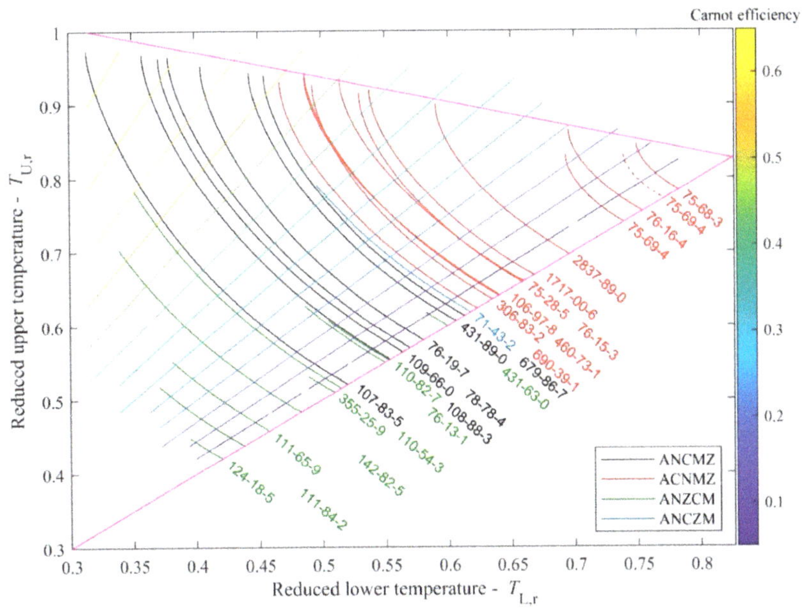

Figure 4. Potential isentropic expansion routes in a reduced T_L-T_U diagram of 28 working fluids, identified by their CAS-numbers. Carnot-efficiencies for a given upper and lower temperature pair are given by color-coded lines.

4. Selection Process of the Thermodynamically Optimal Working Fluid

Concerning an existing heat reservoir, one has to use a fixed maximal or upper temperature (T_U) for the cycle, given by the heat source temperature (T_{source}) and the pinch temperature (ΔT, a characteristic value for the heat exchanger) as:

$$T_U = T_{source} - \Delta T \quad (1)$$

In a similar manner, the lowest temperature of the cycle (T_L) can be defined by the temperature of the environment used for cooling ($T_{environment}$) and a lower pinch temperature (not necessarily equal with the other one) as:

$$T_L = T_{environment} + \Delta T \quad (2)$$

This (T_U; T_L) pair can be used as a characteristic point in Figure 2 to represent a given heat source and sink with other technical details such as quality of heat exchangers; since for reduced quantities, the critical temperature of the working fluid is also necessary, therefore on reduced diagrams (like Figure 3), the reduced (T_U; T_L) points alone cannot be used to characterize the heat source and heat sink of the system.

In Figure 5d, the grey cross represents a hypothetical heat source and sink characterized by T_U = 293 K = 19.85 °C and T_L = 393 K = 119.85 °C. This can be for example a geothermal heat reservoir with a natural cold water source (a small river) or cold internal air (winter) used for cooling. This situation is very similar to the first (and presently only) ORC-based Hungarian geothermal power plant (see the ORC Word Map [22]).

In Figure 5d, only part of the data presented in Figure 2 can be seen, centered on the (293K, 393 K) point. This point is located between two lines representing two working fluids, namely benzene (CAS 71–43–2) and toluene (CAS 108–88–3); one more material can be seen here, namely the 1,1–dichloro–1–fluoroethane (CAS 1717-00-6, also referred as R141b). Since none of these lines run through the 293–392 K point, the closest ones (benzene and toluene) should be considered as nearest to ideal. For the sake of completeness, the potential use of the third one (R141b) is also discussed. The upper and lower temperatures determine the extent of expansion, as it can be seen in Figure 5a,e (benzene), Figure 5b,f (toluene) and Figure 5c,g (R141b).

For further discussion, one can define the so-called quality (q), also referred as dryness or dryness fraction of pure fluids, representing the ratio of vapour within the two-phase region and defined as:

$$q = \frac{n^v}{n^l + n^v} \quad (3)$$

where n^v and n^l are the mole- (or alternatively, the mass-) fraction of the fluid in the vapour or liquid phase, respectively. This quality can change between one (saturated vapour) and zero (saturated liquid); outside of the two-phase region, it is usually not defined or defined as one (vapour states) or zero (liquid states). This ratio can be also calculated by using the specific entropy values for the given mixed state and comparing it with the entropy values of the saturated liquid and vapour states at the same temperature [23].

In the case of benzene (Figure 5a,e), one can see that the expansion route started at saturated vapour state at 393 K and terminated at 293 K, runs in superheated dry vapour region for a while, but the final part of the expansion (starting from 309.96 K down to 293 K, i.e., in an almost 17 K temperature range) runs into the two-phase, wet region. The maximal q value (corresponding to the final state of the expansion) is 0.9848, equal to 1.52% (in mass) of droplet. Droplets may cause erosion problems, moreover they might decrease the net efficiency due to moisture loss, although this 1.52% is not very high compared to the 8–12% which is usually considered tolerable [24]. For benzene, the ideal case—i.e., expansion from one saturated vapour state to another saturated vapour state—would be the termination of expansion at T_L = 309.96 K, given by the cross section of the blue line (representing benzene) and grey dashed line (representing T_U = 393 K). Stopping at this higher temperature would cause smaller efficiency, therefore the problem of droplet erosion should be weighed against the problem of potential efficiency loss, before deciding the use or omission of benzene. It also means that benzene would be the ideal working fluid (concerning only the expansion route and neglecting other selection criteria) for a heat source and sink characterized by (T_U, T_L) = (393 K, 309.96 K) temperatures.

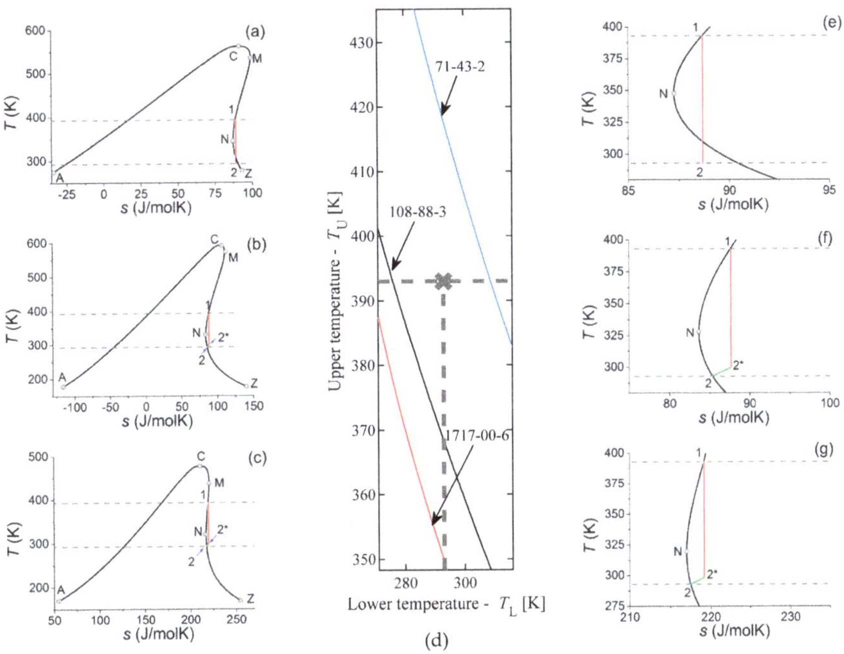

Figure 5. Expansion process of benzene, CAS 71-43-2 (**a**,**e**), toluene, CAS 108-88-3 (**b**,**f**) and 1,1-dichloro-1-fluoroethane, also referred as R141b or CAS 1717-00-6 (**c**,**g**) for a given T_L-T_U pair (293-393 K). Magnified part of Figure 2 centered on the 293-393 K point can be seen in the middle (**d**), representing benzene (blue), toluene (black) and R141b (red). Right-side figures (**e**–**g**) show the magnified parts of (**a**–**c**) around point N, for the better visibility of the expansion lines.

In case of toluene (Figure 5b,f) if the expansion route starts at saturated vapour state at 393 K and terminates at 293 K, than even the final part of the expansion runs in superheated dry vapour states. Therefore one would not face challenges regarding droplet formation. Unfortunately, another problem emerges, namely the problem of cooling down the expanded dry vapour in an isobaric way before reaching saturated vapour state at T_L = 293 K. Therefore, isentropic expansion (solid red line) has to be stopped at a higher temperature (point 2*, T = 299.7 K) and the rest of the process (2*→2) has to proceed on an isobaric cooling route (green line). In this way, a small, almost triangle shaped part in the *T*-*s* diagram (formed by the grey line, the green line and by the extension of the red one) will be excluded from the cycle, decreasing the thermal efficiency. As an additional problem, this unused (lost) heat can increase the heat-load of the environment. Addition of a recuperative heat exchanger to the system can solve—or at least minimize—these problems, since part of this heat can be used to pre-heat the compressed liquid. This solution is appropriate only if the temperature difference between point 2* and 2 is large enough. In case of toluene with 393–293 K expansion, this gap is around ΔT = 7 K, which can be used only in a very limited manner for this purpose. With a larger gap, this heat can be utilized better, however the thermal efficiency is still smaller, therefore the goal remains to reach a perfect "saturated steam to saturated steam" expansion route, with zero gap. Similarly to the case of benzene, one can define an ideal T_L, where expansion from T_U would be terminated at the saturated vapour state; for toluene it would be 276.06 K. It means, that toluene would be the ideal working fluid (concerning expansion route) for a heat source and sink characterized by (T_U, T_L) = (393 K, 276.06 K) temperatures.

In case of R141b (Figure 5c,g) the situation is very similar to the one observed for toluene. The expansion—even in the final stage—runs in superheated dry vapour region, so droplet-related problems do not exist. Similarly to toluene, isentropic expansion has to be stopped at a temperature above 293 K, namely at the one marked as 2* (here it means 298.5 K) and the rest of the process (2*→2) has to proceed on an isobaric cooling route, losing some part of thermal efficiency and increasing the heat-load of the environment. Here, the temperature gap between points 2* and 2 is 5.5 K, which is probably too small to use recuperator. Similarly to the case of benzene and toluene, one can define an ideal T_L, where expansion from T_U would be terminated at the saturated vapour state; for R141b it would be 268.2 K, almost 8 K lower than for toluene. It means, that R141b would be the ideal working fluid (concerning expansion route) for a heat source and sink characterized by (T_U, T_L) = (393 K, 268.2 K) temperatures. One should realize, that for materials on the dry side, but "farther" from the point characterizing the heat source and heat sink (here, toluene is closer, R141b is farther), the expansion has to be stopped farther on temperature scale from the saturated liquid state, but this temperature distance does not correlate with the temperature difference between points 2* and 2. For R141b, the "belly" of the inverse S-shape is smaller than for toluene, therefore the temperature gap between points 2* and 2 is smaller for R141b than for toluene, despite the fact that R141b is farther from the characteristic point. Therefore, R141b might be better for this given heat source and heat sink system, than for toluene, i.e., not only the immediate, but second (or even further) neighbours should be checked in the selection process.

Table 2 summarizes the data which compares the efficiencies of cycles using various materials. Details of the calculations are given in the Supplementary Materials. For all materials, the total efficiencies are in the 21.05–21.72%, and the net efficiencies are in the 20.79–21.66% range. With such small differences—especially for benzene and toluene—the difference of operational (including maintenance) and investment costs by using the simplest layout can be a decisive factor by choosing working fluids. In this case—since the ideal fluid has not been found—the cost related to the presence of droplets has to be compared with the costs related to the installation and operation of the recuperator. Since the use of recuperator are not justified here, toluene seems to be the better choice in the given temperature range; its net efficiency matches the efficiency of benzene but droplets would not be present to cause erosion-related problems. Obviously, using other parameters for the optimization process, other fluids might perform better.

Table 2. Total (η_{br}) and net (η_{net}) cycle efficiencies—calculated in CycleTempo 5.0 environment using fluid data from RefProp 9.0—of ORCs using working fluids presented in Figures 5 and 6. For net efficiencies, self-consumption of pumps were deducted. Temperatures are shown here not only in K, but also in Celsius, according to engineering practice. Values for qualities at the end of expansion are also shown (q). Details of calculations can be seen in Supplementary materials.

T_L	19.85 °C 293 K	19.85 °C 293 K	19.85 °C 293 K	−44.65 °C 228.5 K	−44.65 °C 228.5 K	−44.65 °C 228.5 K
T_U	119.85 °C 393 K	119.85 °C 393 K	119.85 °C 393 K	29.85 °C 303 K	29.85 °C 303 K	29.85 °C 303 K
material	benzene	toluene	R141b	pentane	isopentane	hexane
CAS	71-43-2	108-88-3	1717-00-6	109-66-0	78-78-4	110-54-3
br	21.72%	21.65%	21.05%	21.23%	21.21%	21.28%
net	21.66%	21.63%	20.79%	21.21%	21.18%	21.27%
P_{pump}	−2.75 kW	−1.26 kW	−12.55 kW	−1.04 kW	−1.46 kW	−0.31 kW
P_{turb}	1000 kW	1000 kW	1000 kW	1000 kW	1000 kW	1000 kW
q	98.48%	100.00%	100.00%	99.52%	100.00%	100.00%

Unfortunately, none of the expansion routes of known working fluids run through the (393 K, 293 K) point, therefore one can choose only something "close enough". Another alternative would be—especially when there are two "close enough" working fluids with chemically similar structure, like toluene and benzene—the application of a mixture. Due to the chemical similarity, the mixture would be probably an almost ideal, zeotropic mixture, which can be used more easily as working fluid in ORC applications than azeotropic ones [8,25,26].

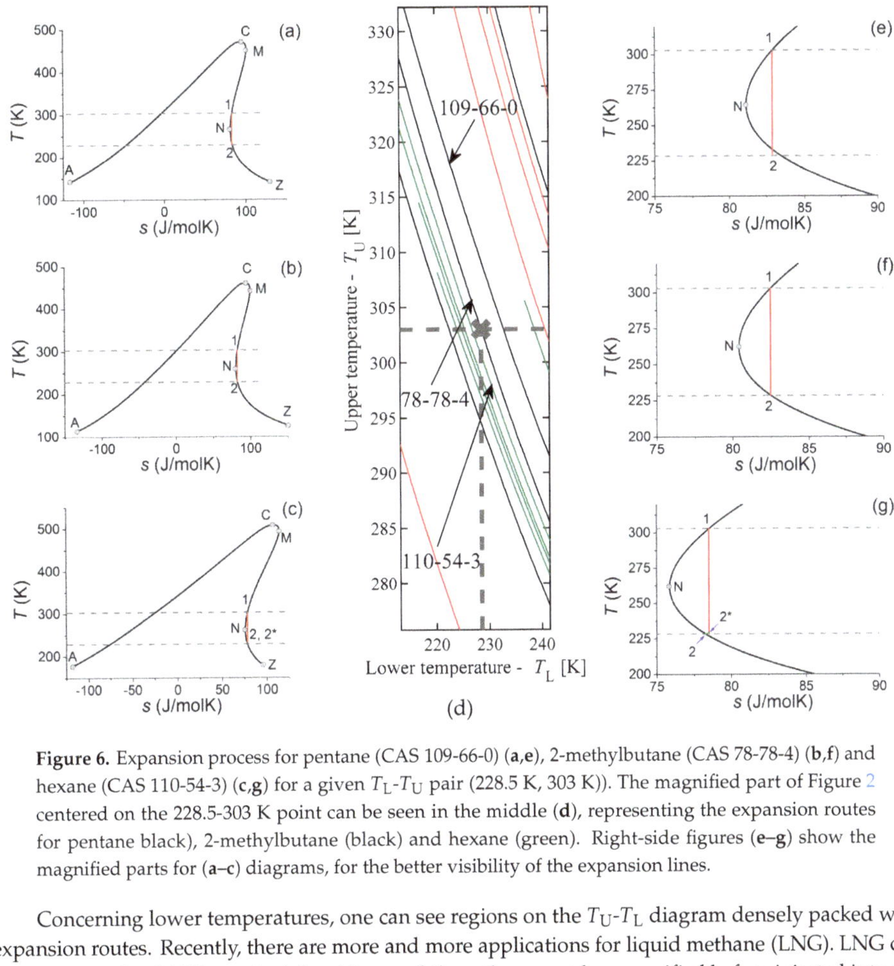

Figure 6. Expansion process for pentane (CAS 109-66-0) (**a**,**e**), 2-methylbutane (CAS 78-78-4) (**b**,**f**) and hexane (CAS 110-54-3) (**c**,**g**) for a given T_L-T_U pair (228.5 K, 303 K)). The magnified part of Figure 2 centered on the 228.5-303 K point can be seen in the middle (**d**), representing the expansion routes for pentane black), 2-methylbutane (black) and hexane (green). Right-side figures (**e–g**) show the magnified parts for (**a–c**) diagrams, for the better visibility of the expansion lines.

Concerning lower temperatures, one can see regions on the T_U-T_L diagram densely packed with expansion routes. Recently, there are more and more applications for liquid methane (LNG). LNG can be transported to LNG-terminals by ships, and there they must be re-gasified before injected into the regular gas network, giving the possibility to utilize the so-called "cold energy" [27,28]. Concerning the fact, that the temperature of liquid methane can be as low as 111 K (−162 °C), the external heat used for re-gasification can be even the environmental heat (air, water), around or below 300 K. Obviously, going down to 111 K in an ORC would cause some serious problems concerning materials used for the construction, but using the methane as heat sink and cooling the chosen working fluid down only to a moderate level, one can define an acceptable expansion range for example between T_U = 303 K and T_L = 228.5 K. Here, the lower temperature is only an arbitrarily chosen, but plausible value to demonstrate the selection process; for any other temperature, the selection process would be the same, only the optimal working fluid would be different. In Figure 6d, one can see part of Figure 2 centered around this point. It is clearly visible, that one line runs almost exactly through this point, namely the line representing 2-methylbutane (isopentane, CAS 78-78-4) and it is slightly off from the lines representing two linear alkanes, hexane (CAS 110-54-3) and pentane (CAS 109-66-0). Expansion lines for these three working fluids are shown in Figure 6a,e (pentane), Figure 6b,f (2-methylbutane) and Figure 6c,g (hexane).

First, the potential use of pentane is discussed. In this case, the end of the expansion runs into the two-phase (wet) region (see Figure 6a,e), but in a much smaller extent, than it was shown for benzene. Here, the expansion line enters the two-phase region around 232.8 K, only 4.3 K above the end of the expansion process. The quality of wet steam will be 0.9952, corresponding to 0.48% wetness, half of the value reported in the previous case. In the case of benzene, the wetness was already in the tolerable level; it is even more valid for this case.

By using 2-methylbutane, the expansion between $T_U = 303$ K and $T_L = 228.5$ K starts and terminates in saturated vapour states, avoiding all problem related to wet or superheated dry vapour. This is shown in Figure 6b,f. Concerning the expansion route, this is the perfect working fluid for a heat source and heat sink pair, represented by $T_U = 303$ K and $T_L = 228.5$ K.

Finally, the use of hexane is discussed. In this case, the expansion should be terminated within the dry vapour region and the working fluid has to be cooled down to T_L by isobaric cooling. Here, the isobaric part is so small, that it can be hardly seen even on the magnified figure (Figure 6c,g); the expansion has to stop at 228.97 K, instead of 228.5 K, only 0.47 K above the T_L. This small gap cannot be utilized for pre-heating the pressurized liquid, but the heat loss caused by this part is negligible.

For these materials, total cycle efficiencies move in the very narrow 21.21–21.28% range, while their net counterparts are in the 21.18–21.27% range. With such small differences, the difference of operational and investment costs by using the most simple layout can be a decisive factor by choosing working fluids; in this case, the most simple layout can be realized by using 2-methylbutane, without the loss of efficiency. Obviously, other parameters of optimization might give a different result, but for our criteria, this is the best-performing working fluid for the given temperature range. It should be noted here, that being the three materials chemically similar, some of their chemistry-related properties (like ODP, GWP, toxicity, flammability) are quite similar, and therefore the indicators related to these properties are also similar.

In the case of $T_U = 303$ K and $T_L = 228.5$ K, one can pinpoint the ideal working fluid, where expansion would run from one saturated vapour state to another saturated vapour state, all along a dry vapour route. At the end of the expansion, the final state would be a saturated vapour state, therefore the problems caused by wet or superheated dry vapour (discussed above) do not exist! This ideal working fluid would be the 2-methylpentane. Due to the fact that expansion routes for several materials run in the immediate vicinity of this $(T_U, T_L) = (228.5$ K, 303 K$)$ point, one can even find a second-, third-, fourth-best; this is especially important, when the working fluid chosen as "best" for our criterion performs worse, when another important optimization criterion is considered. Because these routes are close to our working point, using one of them—instead of the ideal one—would cause only minor problems, acceptable for the given application. For this temperature range, the ones next to the best would be pentane and hexane; using them the problems caused by the presence of droplets or superheated dry vapour are almost negligible, therefore their application (and probably the application of several other working fluids with expansion range lines running close to the $(T_U, T_L) = (228.5$ K, 303 K$)$) would be acceptable. In this way, one could use other criteria (environmental safety, chemical compatibility, price, availability, etc.) together with some optimization method (see for example [29–31]) to choose the proper working fluid, still acceptable thermodynamically, but not necessarily the best one according to this criterion.

5. Conclusions

A novel method has been proposed to choose the best working fluid—from the point of view of expansion route—for a given heat source and heat sink (characterized by a maximum and minimum temperature). The method is based on the novel classification of working fluids using the sequences of their characteristic points on temperature-entropy space [11]. Using this simple method, a possible expansion route can be defined between an upper and lower temperature values; these temperatures are characterizing the given heat source, heat sink and the set of heat exchangers. In this way, one can select a working fluid (from a database, see [14]), where an ideal adiabatic (isentropic) expansion process between a given upper and lower temperature is possible in a way, that the initial and final states are both saturated vapour states and the ideal (isentropic) expansion line runs in the superheated dry vapour region all along the expansion. In this way, it is possible to avoid problems related to the presence of droplets or superheated dry steam in the final expansion state, just before the condensation. Therefore, one can use the most simple ORC layout, using only a pump, two heat exchangers (evaporator, condenser) and an expander, avoiding the use of superheater (or droplet separator) and recuperator; since the fluid is always in dry condition during expansion, droplet erosion can also be avoided.

Study of potential expansion routes in van der Waals model fluids and comparison of the results with the one obtained with real working fluids can help us to single out erroneous data of working fluids presented in various databases.

Presently we have a database of nearly 30 pure fluids with *T-s* data taken from the NIST Chemistry WebBook [15] and from RefProp 9.1 [16]. Most of these fluids were termed formerly as "dry", while in the novel classification they are in various isentropic sub-classes, namely in ANCMZ, ACNMZ, ANZCM and ANCZM. Due to the fact, that the characteristic point responsible for the existence of these expansion routes (point N) is usually located at low temperature for most of the materials of our database, presently this method with the existing, highly accurate database can mostly be used to choose the working fluids thermodynamically most suitable for cryogenic cycles (applied for example for heat recovery during LNG-regasification, see for example [27,28], although some of the materials can be used in higher temperatures (covering part of the temperature range for geothermal and waste heat utilization).

Extension of the method (including internal efficiencies below 1 for the expander as well as acceptable wetness for the final vapour state) and expansion of the database with more material are in progress. After a proper extension of the database, this method will be available to choose ideal working fluids for other temperatures (like the ones characteristic for various geothermal or waste heat applications).

Supplementary Materials: The following are available online at http://www.mdpi.com/1996-1073/12/10/2028/s1, "Cycle efficiencies of ORCs using working fluids presented in Figures 5 and 6".

Author Contributions: Conceptualization, A.R.I.; analysis: R.K. and A.G; writing: A.R.I., A.G. and R.K.

Funding: This work was performed in the frame of FIEK_16-1-2016-0007 project, implemented with the support provided from the National Research, Development and Innovation Fund of Hungary, financed under the FIEK_16 funding scheme. Some part of the research reported in this paper was supported by the Higher Education Excellence Program of the Ministry of Human Capacities in the frame of Nanotechnology research area of Budapest University of Technology and Economics (BME FIKP-NANO) Partial financial supports of the Hungarian National Innovation Office grant (NKFIH, grant No. K116375) is also acknowledged

Conflicts of Interest: The Authors declare no conflict of interest.

References

1. Macchi, E.; Astolfi, M. *Organic Rankine Cycle (ORC) Power Systems: Technologies and Applications*; Elsevier-Woodhead Publishing: Duxford, UK, 2016.
2. Blaise, M.; Feidt, M. Waste Heat Recovery and Conversion into Electricity: Current Solutions and Assessment. *Int. J. Thermodyn.* **2019**, *22*, 1–7. [CrossRef]
3. Chen, H.; Goswami, D.Y.; Stefanakos, E.K. A review of thermodynamic cycles and working fluids for the conversion of low–grade heat. *Renew. Sustain. Energy Rev.* **2010**, *14*, 3059–3067. [CrossRef]

4. Bao, J.; Zhao, L. A review of working fluid and expander selections for Organic Rankine Cycle. *Renew. Sustain. Energy Rev.* **2013**, *24*, 325–342. [CrossRef]
5. Zhang, X.; He, M.; Wang, J. A new method used to evaluate organic working fluids. *Energy* **2014**, *67*, 363–369. [CrossRef]
6. Xia, X.X.; Wang, Z.Q.; Hu, Y.H.; Zhou, N.J. A novel comprehensive evaluation methodology of organic Rankine cycle for parameters design and working fluid selection. *Appl. Therm. Eng.* **2018**, *143*, 283–292. [CrossRef]
7. Garrido, J.M.; Quinteros-Lama, H.; Mejía, A.; Wisniak, J.; Segura, S. A rigorous approach for predicting the slope and curvature of the temperature–entropy saturation boundary of pure fluids. *Energy* **2012**, *45*, 888–899. [CrossRef]
8. Groniewsky, A.; Györke, G.; Imre, A.R. Description of wet-to-dry transition in model ORC working fluids. *Appl. Therm. Eng.* **2017**, *125*, 963–971. [CrossRef]
9. Albornoz, J.; Mejía, A.; Quinteros-Lama, H.; Garrido, J.M. A rigorous and accurate approach for predicting the wet–to–dry transition for working mixtures in organic Rankine cycles. *Energy* **2018**, *156*, 509–519. [CrossRef]
10. Groniewsky, A.; Imre, A.R. Prediction of the ORC working fluid's temperature-entropy saturation boundary using Redlich-Kwong equation of state. *Entropy* **2018**, *20*, 93. [CrossRef]
11. Györke, G.; Deiters, U.K.; Groniewsky, A.; Lassu, I.; Imre, A.R. Novel Classification of Pure Working Fluids for Organic Rankine Cycle. *Energy* **2018**, *145*, 288–300. [CrossRef]
12. White, J.A.; Velasco, S. Characterizing wet and dry fluids in temperature-entropy diagrams. *Energy* **2018**, *154*, 269–276. [CrossRef]
13. White, J.A.; Velasco, S. A Simple Semiempirical Method for Predicting the Temperature−Entropy Saturation Curve of Pure Fluids. *Ind. Eng. Chem. Res.* **2019**, *58*, 1038–1043. [CrossRef]
14. Györke, G.; Groniewsky, A.; Imre, A.R. A simple method to find new dry and isentropic working fluids for Organic Rankine Cycle. *Energies* **2019**, *12*, 480. [CrossRef]
15. NIST Chemistry WebBook, NIST Standard Reference Database Number 69. 2018. Available online: http://webbook.nist.gov/chemistry/ (accessed on 1 December 2018).
16. RefProp 9.1. 2013. Available online: https://www.nist.gov/srd/refprop (accessed on 1 December 2018).
17. Imre, A.R.; Groniewsky, A. Various ways of adiabatic expansion in Organic Rankine Cycle (ORC) and in Trilateral Flash Cycle (TFC). *Z. Phys. Chem.* **2019**, *233*, 577–594. [CrossRef]
18. FluidProp version 3.0.4. 2016. Available online: http://www.asimptote.nl/software/fluidprop/ (accessed on 1 December 2018).
19. Reynolds, W.C. *Thermodynamic Properties in S.I.: Graphs, Tables, and Computational Equations for Forty Substances*; Department of Mechanical Engineering, Stanford University: Stanford, CA, USA, 1979.
20. Imre, A.R.; Groniewsky, A.; Györke, G. Description of the metastable liquid region with quantic and quasi-quintic equation of states. *Interfacial Phenom. Heat Transf.* **2017**, *5*, 173–185. [CrossRef]
21. Taft, R.; Stareck, J. Relationship between Melting-Points, Normal Boiling-Points and Critical Temperatures. *J. Phys. Chem.* **1930**, *34*, 2307–2317. [CrossRef]
22. ORC World Map. 2019. Available online: https://orc-world-map.org/ (accessed on 1 December 2018).
23. Imre, A.R.; Quinones-Cisneros, S.E.; Deiters, U.K. Adiabatic processes in the liquid–vapor two–phase region—1. Pure fluids. *Ind. Eng. Chem. Res.* **2014**, *53*, 13529–13542. [CrossRef]
24. Bassily, A.M. Modeling, numerical optimization, and irreversibility reduction of a dual-pressure heat combined-cycle. *Appl. Energy* **2005**, *81*, 127–151. [CrossRef]
25. Su, W.; Hwang, Y.; Deng, S.; Zhao, L.; Zhao, D. Thermodynamic performance comparison of Organic Rankine Cycle between zeotropic mixtures and pure fluids under open heat source. *Energy Convers. Manag.* **2018**, *165*, 720–737. [CrossRef]
26. Varga, Z.; Csaba, T. Techno-economic evaluation of waste heat recovery by organic Rankine cycle using pure light hydrocarbons and their mixtures as working fluid in a crude oil refinery. *Energy Convers. Manag.* **2018**, *174*, 793–801. [CrossRef]
27. Sadaghiani, M.S.; Ahmadi, M.H.; Mehrpooya, M.; Pourfayaz, P.; Feidt, M. Process development and thermodynamic analysis of a novel power generation plant driven by geothermal energy with liquefied natural gas as its heat sink. *Appl. Therm. Eng.* **2018**, *133*, 645–658. [CrossRef]
28. Atienza-Márquez, A.; Bruno, J.C.; Akisawa, A.; Nakayama, M.; Coronas, A. Fluids selection and performance analysis of a polygeneration plant with exergy recovery from LNG regasification. *Energy* **2019**, *176*, 1020–1036.

29. Quoilin, S.; Van Den Broek, M.; Declaye, S.; Dewallef, P.; Lemort, V. Technoeconomic survey of Organic Rankine Cycle (ORC) systems. *Renew. Sustain. Energy Rev.* **2012**, *22*, 168–186. [CrossRef]
30. Zhang, X.; Zhang, C.; He, M.; Wang, J. Selection and Evaluation of Dry and Isentropic Organic Working Fluids Used in Organic Rankine Cycle Based on the Turning Point on Their Saturated Vapor Curves. *J. Therm. Sci.* **2019**, *28*. [CrossRef]
31. Shi, W.; Pan, L. Optimization Study on Fluids for the Gravity-Driven Organic Power Cycle. *Energies* **2019**, *12*, 732. [CrossRef]

energies

Article

Approximating the Temperature–Entropy Saturation Curve of ORC Working Fluids From the Ideal Gas Isobaric Heat Capacity

Juan A. White [1,2,*,†] and Santiago Velasco [1,2,†]

1 Departamento de Física Aplicada, Universidad de Salamanca, 37008 Salamanca, Spain
2 IUFFyM, Universidad de Salamanca, 37008 Salamanca, Spain
* Correspondence: white@usal.es
† These authors contributed equally to this work.

Received: 29 July 2019; Accepted: 21 August 2019; Published: 24 August 2019

Abstract: Recently, we proposed an approximate expression for the liquid–vapor saturation curves of pure fluids in a temperature–entropy diagram that requires the use of parameters related to the molar heat capacity along the vapor branch of the saturation curve. In the present work, we establish a connection between these parameters and the ideal-gas isobaric molar heat capacity. The resulting new approximation yields good results for most working fluids in Organic Rankine Cycles, improving the previous approximation for very dry fluids. The ideal-gas isobaric molar heat capacity can be obtained from most Thermophysical Properties databases for a very large number of substances for which the present approximation scheme can be applied.

Keywords: ORC working fluids; temperature–entropy saturation curve; saturation properties; wet and dry fluids; ideal-gas heat capacity

1. Introduction

There is an increasing interest in the use of Organic Rankine Cycles (ORCs) as a suitable way of generating power from low-temperature heat sources such as geothermal, solar thermal, biomass, waste heat, and bottoming cycles. As is well known, a key aspect in the optimal implementation of an ORC for a given heat source is the choice of the working fluid. An appropriate working fluid selection should take into account several criteria such as thermo-economic efficiency, safety, environmental aspects, chemical stability, etc. [1–6]

One of the most relevant aspects in ORC working fluid selection is the analysis of the shape of the liquid–vapor saturation curve in a temperature–molar entropy (T-s) diagram because it has a direct influence both in the thermal efficiency and in the particular design of the cycle. Let us consider a simple, ideal ORC with an evaporation temperature $T_{\rm ev}$ and a condensation temperature $T_{\rm con}$, so that $T_{\rm con} < T_{\rm ev} < T_c$ where T_c is the critical temperature. In this simple ORC, the isentropic expansion that takes place in the turbine and starts from a saturated vapor state at $T_{\rm ev}$ can lead to three different situations depending on the shape of the saturation curve: (1) If the mean slope of the vapor branch of the T-s saturation curve between $T_{\rm con}$ and $T_{\rm ev}$ is negative, the working fluid has a *wet fluid behavior* and the isentropic expansion process in the turbine gives rise to condensation, i.e., it ends in the two-phase region of the T-s diagram. This situation should be avoided (via superheating) since the mixture of vapor with liquid droplets could lead to damage of the turbine blades [7]. (2) If the mean slope is positive, the working fluid has a *dry fluid behavior* and the isentropic expansion leads to superheated vapor. This implies a reduction in the cycle efficiency which can only be partially remediated by resorting to a regenerator. (3) Finally, if the absolute value of the mean slope is very large, the working fluid *behaves as an isentropic fluid* so that the turbine isentropic expansion ends near the saturated vapor state at T_{con}. In this case, neither regeneration nor superheating is required.

Fluids such as water, carbon dioxide or ammonia that always have a negative slope in the vapor branch of the T-s saturation curve ($dT/ds_g < 0$) always *behave as wet fluids* and are usually termed as *wet*. On the other hand, fluids such as siloxanes or alkanes with a large number of carbons have a positive slope for most temperatures in the T-s saturation curve and are called *dry fluids* since they usually lead to a *dry fluid behavior*. Finally, fluids like the refrigerants RE143a, R11, or R116 present a wide range of temperatures for which the saturated vapor curve is almost vertical. These fluids are usually termed as *isentropic*, although *isentropic behavior* can also be obtained with a particular class of *dry fluids* [8–10]. To summarize, the slope of the vapor branch of the T-s saturation curve gives rise to a basic classification of working fluids into three categories: wet, dry, and isentropic. According to Liu et al. dry or isentropic fluids are preferred for ORC applications since they eliminate the problems related to condensation in the isentropic expansion [7].

Most studies on the shape of the T-s saturation boundary have focused in the analysis of the slope of the vapor branch dT/ds_g and its relation with the molar heat capacity of the fluid. In a seminal work, Morrison [11] presented a study in the context of refrigeration cycles, concluding that negative slopes in the vapor branch do arise for fluids with large isochoric molar heat capacity and, consequently, for fluids with large, complex molecules. Liu et al. [7] derived an approximate expression of the (inverse) slope $\xi = ds_g/dT$ in terms of the isobaric molar heat capacity of the saturated vapor and the molar enthalpy of vaporization, and related the dry or wet character of the fluid to the sign of ξ. Other authors have obtained results for different model equations of state, concluding that a key aspect in the shape of the T-s saturation curve is the ideal gas contribution to the molar heat capacity of the fluid [12–15].

Recently [16], we proposed a semiempirical method to obtain the T-s saturation curve by means of: (1) a modified rectilinear diameter law for the two branches of the saturated entropy; and (2) an appropriate expression for the entropy of vaporization Δs. The method only requires the knowledge of three parameters: T_c, the acentric factor ω, and the slope of the modified rectilinear diameter $-b$. Two approximations were considered for b. The simplest one requires the knowledge of the critical molar volume v_c, which together with T_c and ω can be obtained in any Thermophysical Properties database. The most accurate approximation requires the calculation of the slope ξ at a certain temperature and therefore, the use of programs like RefProp [17] or CoolProp [18]. The goal of the present work is to study the relation between the modified rectilinear diameter law and the ideal gas molar heat capacity considered in earlier works [12–15]. More concretely, in this work, we propose a new approximation for b in terms of ω and the ideal-gas isobaric molar heat capacity at an appropriate reduced temperature. The new approximation has the same application range as the previous ones, with the advantage that it does not require the calculation of ξ and yields better results for very dry fluids.

This paper is structured as follows. In Section 2, we outline the main features of the semiempirical method in [16]. In Section 3, we introduce the ideal gas contribution to the entropy and investigate its relevance in the dry or wet behavior of a working fluid. From the results of the preceding sections, a new approximation for the slope of the modified rectilinear diameter is proposed in Section 4, in terms of the ideal gas molar heat capacity of the fluid. In Section 5, we compare the results of the new approximation with previous ones. Finally, we conclude in Section 6 with a brief summary.

2. A Semiempirical Method for the T-s Saturation Curve

In [16], we proposed a semiempirical method to obtain approximate expressions for the liquid–vapor phase boundary in a temperature–entropy diagram. Here, we present the basic equations of the method that are employed in connection with the ideal gas molar heat capacity of the fluid. We find it convenient to work in reduced coordinates so that, without loss of generality, we consider a $T_r - s^*$ diagram where $T_r = T/T_c$ is the reduced temperature and $s^* = (s - s_c)/R$

is a dimensionless molar entropy, where s_c is the molar critical entropy and R is the gas constant. In this context, the entropies of the gas and liquid branches of the phase boundary are denoted as $s_{\rm g}^*$ and $s_{\rm l}^*$, respectively, and are functions of T_r. Furthermore, with these definitions, for all fluids, one has $s_{\rm g}^*(T_r \to 1) = s_{\rm l}^*(T_r \to 1) = 0$.

To determine $s_{\rm g}^*(T_r)$ and $s_{\rm l}^*(T_r)$, two relations are considered in the semiempirical method. The first one is the well-known relation between the dimensionless molar enthalpy of vaporization $\Delta_{\rm v}h_r = \Delta_{\rm v}h/RT_c$ and the molar entropy of vaporization $\Delta_{\rm v}s^* = s_{\rm g}^* - s_{\rm l}^*$:

$$\Delta_{\rm v}h_r = T_r \Delta_{\rm v}s^* \,. \tag{1}$$

Several approximate expressions for $\Delta_{\rm v}h_r$ are available in the literature [19]. For simplicity, we consider the following corresponding states version [20] of the Watson equation [21]:

$$\Delta_{\rm v}h_r = K(\omega)(1 - T_r)^{0.38} \,, \tag{2}$$

where

$$K(\omega) = 7.2729 + 10.4962\,\omega + 0.6061\,\omega^2 \tag{3}$$

is a function of the acentric factor of the fluid ω.

The second relation in the semiempirical method is based on the observation [16] that the line of constant quality q in the two-phase region of the $T_r - s^*$ diagram,

$$s_q^*(T_r) = q s_{\rm g}^*(T_r) + (1-q) s_{\rm l}^*(T_r) \,, \tag{4}$$

has an approximately linear behavior in the range $0.6 < T_r < 0.99$ for an appropriate value of q. The optimal value of q varies slightly with the fluid, with a mean value $\bar{q} = 0.385$ for the 121 fluids of the RefProp 9.1 program [17]. Therefore, one can write the following *modified rectilinear diameter* relation for the entropies of the saturated liquid and vapor [16]:

$$s_{\bar{q}}^*(T_r) \approx b(1 - T_r) \qquad (0.6 < T_r < 0.99) \,, \tag{5}$$

where $-b$ is the slope of the modified rectilinear diameter. We note here that this linear relation is similar to the well known rectilinear diameter law of Cailletet and Mathias [22] for the saturation densities but, instead of using $q = 0.5$, one has to consider a different value of q for the entropy.

Equations (1) and (4) allow expressing the saturation entropies in terms of $\Delta_{\rm v}h_r$ and $s_q^*(T_r)$:

$$s_{\rm g}^*(T_r) = s_q^*(T_r) + (1-q)\frac{\Delta_{\rm v}h_r}{T_r} \,, \tag{6}$$

$$s_{\rm l}^*(T_r) = s_q^*(T_r) - q\frac{\Delta_{\rm v}h_r}{T_r} \,, \tag{7}$$

that are exact relations valid for any value of the quality in the range $0 \le q \le 1$. Substituting the approximations Equations (2) and (5) with $q = \bar{q} = 0.385$, it is direct to obtain the approximate expressions

$$s_{\rm g}^*(T_r) = b(1 - T_r) + (1-\bar{q})K(\omega)\frac{(1-T_r)^{0.38}}{T_r} \,, \tag{8}$$

$$s_{\rm l}^*(T_r) = b(1 - T_r) - \bar{q}K(\omega)\frac{(1-T_r)^{0.38}}{T_r} \,, \tag{9}$$

which are expected to yield good results for the entropies of the saturated vapor and liquid in the range $0.6 < T_r < 0.99$. Equations (8) and (9) require the parameter b as an input. Differentiating Equation (8) with respect to T_r one has

$$\frac{\mathrm{d}s^*_{\mathrm{g}}}{\mathrm{d}T_r} = -b - (1-\bar{q})K(\omega)\frac{(1-0.62T_r)}{T_r^2(1-T_r)^{0.62}} . \tag{10}$$

From this equation, two approximations were considered in [16] for b. The first one, referred to as approximation A1, is given by the following expression:

$$b = -\xi^*_{\mathrm{M}} - (1-\bar{q})K(\omega)\frac{(1-0.62T_{\mathrm{M}r})}{T_{\mathrm{M}r}^2(1-T_{\mathrm{M}r})^{0.62}} , \tag{11}$$

where $T_{\mathrm{M}r}$ is the reduced temperature at which the derivative $\mathrm{d}s^*_{\mathrm{g}}/\mathrm{d}T_r$ attains its maximum value ξ^*_{M} (see [16,23] for details). The values of $T_{\mathrm{M}r}$ and ξ^*_{M} have been calculated in [23] for the 121 pure fluids of RefProp 9.1. From now on, all results obtained from this approximation are labeled with the subscript A1, i.e., we refer to b_{A1}, $s^*_{\mathrm{g,A1}}$, $s^*_{\mathrm{l,A1}}$, and $s^*_{\bar{q},\mathrm{A1}}$.

To avoid the dependence of the parameter b on $T_{\mathrm{M}r}$ and ξ^*_{M}, a further approximation A2 was presented in [16], in which $T_{\mathrm{M}r}$ was replaced by its mean value $\overline{T}_{\mathrm{M}r} = 0.81$ and a correlation in terms of the critical molar volume v_c was used for ξ^*_{M}. Approximation A1 is more accurate, but approximation A2 only depends on T_c, ω, and v_c that are easily accessible for several fluids.

3. The Ideal Gas Contribution to the Entropy

The dimensionless molar entropy of a fluid can be expressed as

$$s^*(T_r, v_r) = s^{\mathrm{ig}*}(T_r, v_r) + s^{\mathrm{r}*}(T_r, v_r) , \tag{12}$$

where $v_r = v/v_c$ is the reduced volume, $s^{\mathrm{r}*} = (s^{\mathrm{r}} - s_c)/R$ is a dimensionless residual entropy (that depends on the equation of state of the fluid), and $s^{\mathrm{ig}*} = s^{\mathrm{ig}}/R$ is the ideal gas contribution to the entropy:

$$s^{\mathrm{ig}*}(T_r, v_r) = \int_{T_{r,0}}^{T_r} \frac{c_p^{\mathrm{ig}*}(T_r) - 1}{T_r}\,\mathrm{d}T_r + \log\left(\frac{v_r}{v_{r,0}}\right) + s_0^{\mathrm{ig}*} . \tag{13}$$

In Equation (13), the subscript 0 refers to an arbitrary reference state and $c_p^{\mathrm{ig}*}(T_r) = c_p^{\mathrm{ig}}(T_r)/R$ is the dimensionless ideal gas isobaric heat capacity. Several approximate expressions for $c_p^{\mathrm{ig}*}$ have been used in the literature. Common choices are polynomial expansions of the form

$$c_p^{\mathrm{ig}*}(T_r) = \sum_{i=0,4} a_i T_r^i , \tag{14}$$

and the well-known Aly-Lee equation [24] used by the DIPPR database [25]

$$c_p^{\mathrm{ig}*}(T_r) = d_0 + d_1\left[\frac{d_2/T_r}{\sinh(d_2/T_r)}\right] + d_3\left[\frac{d_4/T_r}{\sinh(d_4/T_r)}\right] , \tag{15}$$

where the parameters d_i (a_i in Equation (14)) are obtained by a fit to experimental data, and are, therefore, fluid-dependent. Furthermore, it is worth mentioning that, depending on the fluid, the RefProp 9.1 program [17] uses different approximate expressions for c_p^{ig} that provide an accurate fit to experimental data.

We find it convenient to split the ideal gas contribution to the entropy in two parts:

$$s^{\mathrm{ig}*}(T_r, v_r) = s^{\mathrm{ig0}*}(T_r) + s^{\mathrm{ig1}*}(T_r, v_r) , \tag{16}$$

where

$$s^{\text{ig0}*}(T_r) = -\int_{T_r}^{1} \frac{c_p^{\text{ig}*}(T_r)}{T_r}\,\mathrm{d}T_r\,, \tag{17}$$

and

$$s^{\text{ig1}*}(T_r, v_r) = \log\left(\frac{T_{r,0} v_r}{T_r v_{r,0}}\right) + s_0^{\text{ig}*} - s^{\text{ig0}*}(T_{r,0}) \tag{18}$$

so that $s^{\text{ig0}*}$ carries the temperature dependence of the dimensionless molar entropy due to $c_p^{\text{ig}*}$, and $s^{\text{ig1}*}$ is a function of the temperature and the molar volume. From Equations (12) and (16)–(18), one can write the following expressions for the entropies of the saturated vapor and liquid

$$s_{\text{g}}^*(T_r) = s^{\text{ig0}*}(T_r) + s^{\text{ex}*}(T_r, v_{r,\text{g}})\,, \tag{19}$$

$$s_{\text{l}}^*(T_r) = s^{\text{ig0}*}(T_r) + s^{\text{ex}*}(T_r, v_{r,\text{l}})\,, \tag{20}$$

where $v_{r,\text{g}}$ and $v_{r,\text{l}}$ are, respectively, the reduced volumes of the saturated vapor and liquid. The excess entropy $s^{\text{ex}*}(T_r, v_r) = s^{\text{ig1}*}(T_r, v_r) + s^{\text{r}*}(T_r, v_r)$, is a function of the reduced temperature and the reduced volume of the fluid. Equations (19) and (20) show that the entropies of the saturated liquid and vapor can be separated in two contributions. The first one, $s^{\text{ig0}*}$, is the same for the two saturated phases and only depends on T_r through an integral of $c_p^{\text{ig}*}/T_r$. The second contribution $s^{\text{ex}*}$ is different for each saturated phase and depends on the equation of state of the fluid.

The behavior of the ideal gas contribution $s^{\text{ig0}*}(T_r)$ defined in Equation (17) is shown in Figure 1 in a $T_\text{r} - s^*$ diagram for: (a) ammonia; (b) benzene; and (c) methyl palmitate. In this figure, one can see that the main source for the inclination of the $T_\text{r} - s^*$ saturation boundary is the ideal gas contribution $s^{\text{ig0}*}(T_r)$. This implies that the dry or wet character of a fluid is mainly driven by the slope of $s^{\text{ig0}*}(T_r)$. This fact was previously observed by Groniewsky and coworkers [14,15], but using the contribution to the molar entropy due to the isochoric molar heat capacity of the ideal gas c_v^{ig} instead of the isobaric one, c_p^{ig}.

We also show in Figure 1 the line of constant quality $s_{\bar{q}}^*$ (with $\bar{q} = 0.385$) obtained from Equation (4) using RefProp 9.1 results [17] for s_l^* and s_g^*, and the result $s_{\bar{q},\text{A1}}^*$ of approximation A1 for the modified rectilinear diameter relation in Equation (5) with the parameter b_{A1} given by Equation (11). To calculate b_{A1}, the following values were used [23] (see also Table A1): $\omega = 0.256$, $T_{\text{M}r} = 0.8162$, and $\xi_\text{M}^* = -8.6109$ for ammonia; $\omega = 0.211$, $T_{\text{M}r} = 0.8252$, and $\xi_\text{M}^* = 6.1558$ for benzene; and $\omega = 0.91$, $T_{\text{M}r} = 0.7388$, and $\xi_\text{M}^* = 86.6279$ for methyl palmitate. We obtained $b_{\text{A1}} = -4.4225$ for ammonia, $b_{\text{A1}} = -18.5302$ for benzene, and $b_{\text{A1}} = -110.949$ for methyl palmitate.

In Figure 1a, we consider ammonia as an example of wet fluid. As one can observe, in this case, the ideal gas contribution $s^{\text{ig0}*}(T_r)$ shows a noticeable deviation from both the line of constant quality $s_{\bar{q}}^*$ obtained from RefProp 9.1 and the approximate (straight line) result $s_{\bar{q},\text{A1}}^*$. The best agreement for ammonia was obtained for $T_r \approx 0.8$. Much better agreement was obtained for benzene (Figure 1b), especially in the range $0.7 < T_r < 0.99$. The case of methyl palmitate presented in Figure 1c shows a good agreement between $s^{\text{ig0}*}(T_r)$ and $s_{\bar{q}}^*$, with larger deviations with $s_{\bar{q},\text{A1}}^*$. We note that methyl palmitate is a "very dry" fluid, with a large value of ξ_M^* and approximation A1 is known to give large deviations for these fluids [16]. In Figure 1d, we plot the results for $s^{\text{ig0}*}$, $s_{\bar{q}}^*$, and $s_{\bar{q},\text{A1}}^*$ in the same scale, showing that the main discrepancies take place between $s_{\bar{q}}^*$ and $s_{\bar{q},\text{A1}}^*$ for methyl palmitate.

From the results presented in Figure 1, which are also valid for the remaining fluids of RefProp 9.1 (not shown), we can conclude that the ideal gas contribution $s^{\text{ig0}*}$ is close to the line of constant quality $s_{\bar{q}}^*$ with $\bar{q} = 0.385$ but some differences do arise, especially for $T_r < 0.7$. Therefore, it is not advisable to replace $s_{\bar{q}}^*$ with $s^{\text{ig0}*}$ in Equations (6) and (7) in order to obtain a new approximation for the entropy of the saturation boundary. As shown in the next section, it is more convenient to consider the approximate expressions of Equations (8) and (9) but using information from $s^{\text{ig0}*}$ instead of Equation (11).

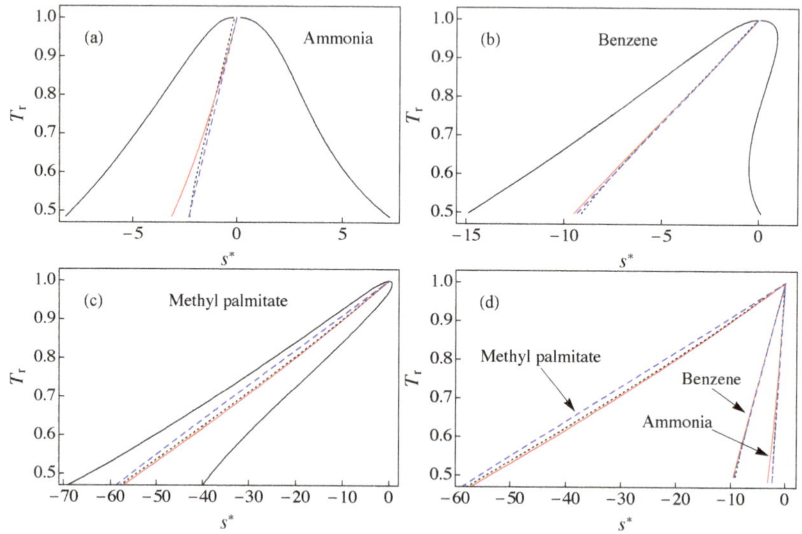

Figure 1. T_r-s^* diagram for: (**a**) ammonia; (**b**) benzene; and (**c**) methyl palmitate. The solid red line corresponds to the ideal gas contribution to the entropy $s^{\mathrm{ig}0*}$ defined in Equation (17), the dotted black line is the constant quality entropy $s^*_{\bar{q}}$ obtained from Equation (4) with $\bar{q} = 0.385$, and the dashed blue line shows $s^*_{\bar{q},\mathrm{A1}} = b(1 - T_r)$ with b obtained from Equation (11) (approximation A1). The solid black line represents the liquid–vapor saturation boundary. (**d**) A comparison of the results for $s^{\mathrm{ig}0*}$, $s^*_{\bar{q}}$, and $s^*_{\bar{q},\mathrm{A1}}$ presented in (**a**–**c**). All data were obtained using RefProp 9.1 [17].

4. A New Approximation for the *T*-*s* Saturation Curve

Differentiating Equation (19) with respect to T_r, we obtain the following expression for the slope of the entropy of the saturated vapor

$$\frac{\mathrm{d}s^*_{\mathrm{g}}}{\mathrm{d}T_r} = \frac{c_p^{\mathrm{ig}*}(T_r)}{T_r} - \frac{\psi^*(T_r)}{T_r}, \tag{21}$$

where

$$\psi^*(T_r) = -T_r \frac{\mathrm{d}s^{\mathrm{ex}*}(T_r, v_{r,\mathrm{g}}(T_r))}{\mathrm{d}T_r}. \tag{22}$$

We note that, as pointed out by Garrido et al. [12], according to Equation (21), a fluid behaves as dry for reduced temperatures such that

$$c_p^{\mathrm{ig}*}(T_r) > \psi^*(T_r). \tag{23}$$

A rigorous approach for obtaining ψ^* from an explicit equation of state (Eos) model has been developed by Garrido et al. [12]. This implies that ψ^* should be correlated with the acentric factor ω of the fluid. In Figure 2, we plot ψ^* vs. ω for the 121 fluids of RefProp 9.1 [17] at a reduced temperature $T_r = \overline{T}_{\mathrm{M}r} = 0.81$, which is taken as reference. Instead of considering a given Eos model, the values of ψ^* were obtained from Equation (21) and RefProp 9.1 data for c_p^{ig} and $\mathrm{d}s^*_{\mathrm{g}}/\mathrm{d}T_r$. As one can observe, there is a rather good correlation between ψ^* and ω. A quadratic fit excluding two oddball fluids (ethanol and methanol, black dots in Figure 2) yields

$$\psi^*(\omega, \overline{T}_{\mathrm{M}r}) = 8.7872 + 8.7191\omega - 1.9704\omega^2 \tag{24}$$

with a coefficient of determination $R^2 = 0.9838$.

Equating Equations (10) and (21) and taking as reference the reduced temperature $\overline{T}_{\mathrm{M}r} = 0.81$, we obtain

$$b_{\mathrm{A3}} = -\frac{c_p^{\mathrm{ig}*}(\overline{T}_{\mathrm{M}r}) + \delta(\omega)}{\overline{T}_{\mathrm{M}r}}, \tag{25}$$

where the label A3 has been chosen to compare with approximations A1 and A2 in [16], and, using Equations (3) and (24),

$$\begin{aligned} \delta(\omega) &= -\psi^*(\omega, \overline{T}_{\mathrm{M}r}) + (1-\bar{q})K(\omega)\frac{(1-0.62\overline{T}_{\mathrm{M}r})}{\overline{T}_{\mathrm{M}r}(1-\overline{T}_{\mathrm{M}r})^{0.62}} \\ &= -1.0901 + 2.3893\omega + 2.6119\omega^2 . \end{aligned} \tag{26}$$

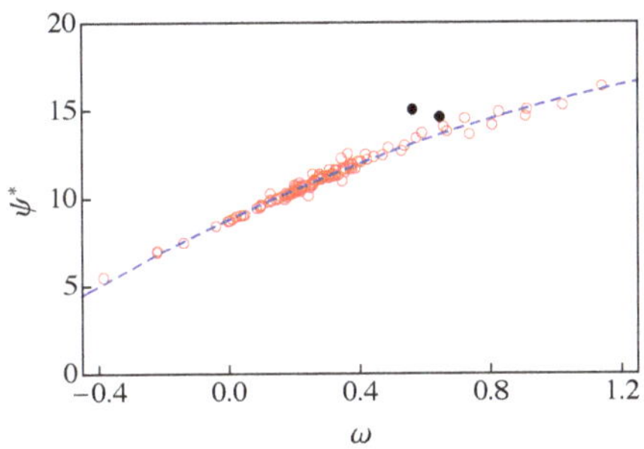

Figure 2. ψ^* vs. ω for the 121 fluids of RefProp 9.1 [17] at a reduced temperature $\overline{T}_{\mathrm{M}r} = 0.81$. The dashed blue line is the result of a quadratic fit to the data (red circles) excluding ethanol and methanol (solid black circles).

Finally, inserting Equation (25) into Equations (8) and (9), we obtain

$$s^*_{\mathrm{g,A3}}(T_r) = -\frac{c_p^{\mathrm{ig}*}(\overline{T}_{\mathrm{M}r}) + \delta(\omega)}{\overline{T}_{\mathrm{M}r}}(1-T_r) + (1-\bar{q})K(\omega)\frac{(1-T_r)^{0.38}}{T_r}, \tag{27}$$

$$s^*_{\mathrm{l,A3}}(T_r) = -\frac{c_p^{\mathrm{ig}*}(\overline{T}_{\mathrm{M}r}) + \delta(\omega)}{\overline{T}_{\mathrm{M}r}}(1-T_r) - \bar{q}K(\omega)\frac{(1-T_r)^{0.38}}{T_r}, \tag{28}$$

which are the results of our new approximation A3 for the entropies of the saturated vapor and liquid. Taking into account that the new approximation is also based in the modified rectilinear diameter relation in Equation (5), we expect that the results of A3 should apply in the range $0.6 < T_r < 0.99$. We would like to recall here that Equations (27) and (28) only require the knowledge of the acentric factor of the fluid, its critical temperature, and the ideal gas molar isobaric heat capacity at a reduced temperature $\overline{T}_{\mathrm{M}r} = 0.81$. These parameters are easily accessible for a large amount of fluids using well known databases such as DIPPR [25].

In the next section, we analyze the results of Equations (27) and (28) by comparing with RefProp data and with the results of approximation A1 derived in [16].

5. Results

Figure 3 compares the results of RefProp 9.1 with those of approximations A1 and A3 for the liquid–vapor saturation boundary in a $T_r - s^*$ diagram for: (a) ammonia; (b) benzene; (c) methyl palmitate;and (d) D6. In the case of ammonia, approximation A3 gave better agreement with RefProp 9.1 than A1 for the saturated vapor branch, whereas we obtained the opposite behavior for the saturated liquid branch. This can be ascribed to the fact that the slope of the modified rectilinear diameter is slightly larger for A3 ($-b_{\mathrm{A3}} = 5.0275$, obtained from Equation (25) with $c_p^{\mathrm{ig}*}(\overline{T}_{\mathrm{M}r}) = 4.3795$) than for

A1 ($-b_{\rm A1} = 4.4225$). Figure 3b shows that A1 and A3 yield similar results for the saturation entropies of benzene, with excellent agreement with RefProp 9.1 results except for $s_{\rm g}^*$ for $T_r < 0.6$. In this case, the slope $-b_{\rm A1} = 18.5302$ is very similar to $-b_{\rm A3} = 18.4997$ (obtained from $c_p^{\rm ig*}(\overline{T}_{\rm Mr}) = 15.4544$). In Figure 3c, the approximation A3 gives a better agreement with RefProp 9.1 than A1. This indicates that the slope $-b_{\rm A3} = 107.9614$ (obtained from $c_p^{\rm ig*}(\overline{T}_{\rm Mr}) = 84.2017$) yields better results than that of approximation A1 ($-b_{\rm A1} = 110.949$). Finally, Figure 3d shows that A3 yields much better agreement with RefProp 9.1 than A1. We note that D6 gives rise to the maximum deviation between A1 and RefProp 9.1, as we shown below. This is clearly due to the fact that the slope $-b_{\rm A1} = 128.0031$ for D6 is much larger than $-b_{\rm A3} = 120.7889$.

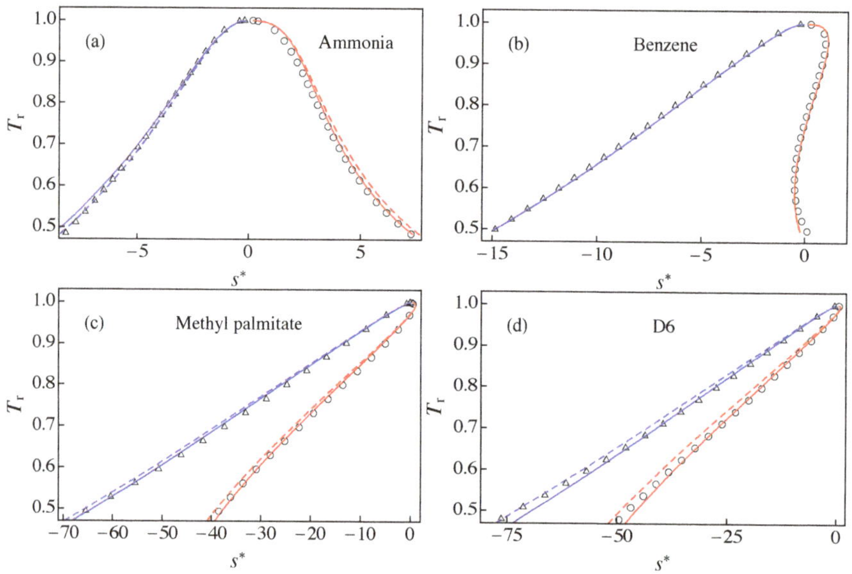

Figure 3. Liquid–vapor saturation boundary in a T_r-s^* diagram for: (**a**) ammonia; (**b**) benzene; (**c**) methyl palmitate; and (**d**) D6. The solid lines correspond to the results of approximation A3 for the entropies of the saturated vapor $s_{\rm g,A3}^*$ (solid red line) and the saturated liquid $s_{\rm l,A3}^*$ (solid blue line). The dashed lines correspond to the results of approximation A1 for the entropies of the saturated vapor $s_{\rm g,A1}^*$ (dashed red line) and the saturated liquid $s_{\rm l,A1}^*$ (dashed blue line). The symbols are RefProp 9.1 results for $s_{\rm g}^*$ (circles) and $s_{\rm l}^*$ (triangles).

The only difference between approximations A1 and A3 comes from the value of the parameter b that is calculated from Equation (11) in the case A1 and from Equation (25) for A3. In Figure 4, we present a plot of $b_{\rm A3}$ vs. $b_{\rm A1}$ for the 121 fluids of RefProp 9.1 [17], which are also listed in Table A1. In all cases, we obtained negative values of b in the range $-130 < b < -1$ (notice that the modified rectilinear diameter relation is given by $b(1 - T_r)$ and thus the slope of the straight line, $-b$, is always positive). Overall, the agreement between both approaches is very good except for very dry fluids (with large values of $-b$) where approximation A3 yields results for $-b$ slightly smaller than those of approximation A1, as one can see in Figure 4a. Wet fluids are those for which $\xi_{\rm M}^* < 0$ and thus they present smaller values of $-b$. As shown in Figure 4b, the agreement between the results of A1 and A3 for wet fluids is still good but presents more spread. The transition from dry to wet fluids occurs for $b \approx -14$ but it is not sharp, mainly due to the effect of the acentric factor in Equation (11).

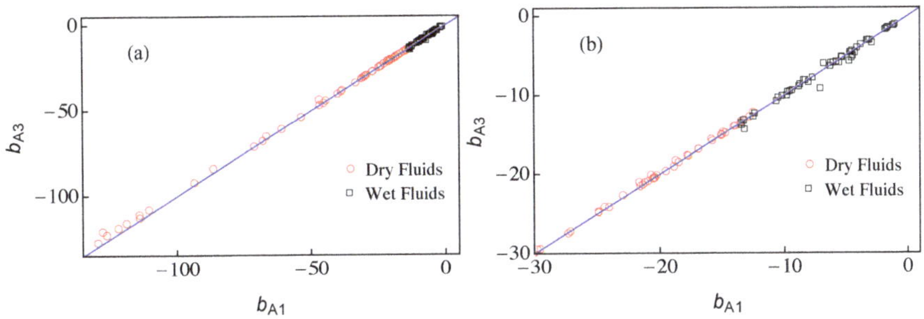

Figure 4. The parameter b_{A3} vs. b_{A1}. The symbols represent the values of b_{A3} and b_{A1} obtained from Equations (25) and (11), respectively, using RefProp 9.1 data [17]. The solid blue line corresponds to the identity $b_{A3} = b_{A1}$. Dry fluids ($\xi_M^* > 0$) are plotted with red circles and wet fluids ($\xi_M^* < 0$) correspond to black squares. (**b**) A zoom of (**a**). The data for b_{A3} and b_{A1} are listed in Table A1.

To provide a quantitative measurement of the performance of approximations A1 and A3, we consider the following expression for the percent relative deviation Δ_{ri} of approximation Ai (i = 1, 3), which was introduced in [16]:

$$\Delta_{ri} = 100\frac{\int_{0.6}^{1}|s_g^*(T_r) - s_{g,Ai}^*(T_r)|dT_r + \int_{0.6}^{1}|s_l^*(T_r) - s_{l,Ai}^*(T_r)|dT_r}{\int_{0.6}^{1}|s_g^*(T_r) - s_l^*(T_r)|dT_r}, \tag{29}$$

where s_g^* and s_l^* are RefProp 9.1 results. We obtained $\Delta_{r1} = 4.82\%$ and $\Delta_{r3} = 3.42\%$ for ammonia, $\Delta_{r1} = 2.11\%$ and $\Delta_{r3} = 2.28\%$ for benzene, and $\Delta_{r1} = 24.68\%$ and $\Delta_{r3} = 14.12\%$ for methyl palmitate.

Figure 5 compares the percent relative deviations of A1 and A3 for the 121 fluids of RefProp 9.1. To relate the performance of A1 and A3 to the dry or wet character of the fluid, the percent relative deviations are plotted as a function of the parameter ξ_M^*. We note that, for the sake of clarity, the results are plotted with lines instead of symbols. As one can observe in Figure 5 (see also Table A1) the new approximation A3 fares better than A1 for fluids with ξ_M^*>28, i.e., for very dry fluids. For the remaining fluids, both approximations yield similar results, in most cases with relative deviations less than 5% (below the dotted line in Figure 5). More concretely, in the case of the new approximation A3, we obtained an average percent relative deviation $\overline{\Delta}_{r3} = 4.19\%$ for all fluids of Table A1, with a maximum percent relative deviation $\max(\Delta_{r3}) = 15.56\%$ for methyl stearate (the only fluid with $\Delta_{r3} > 15\%$). In the case of A1, one has [16] $\overline{\Delta}_{r1} = 5.00\%$ and $\max(\Delta_{r1}) = 34.10\%$ for D6, with nine fluids of a total of 121 with $\Delta_{r1} > 15\%$. The reason for obtaining $\overline{\Delta}_{r1} > \overline{\Delta}_{r3}$ can be clearly ascribed to the poorer behavior of A1 for very dry fluids. We note, however, that, in the two approximate treatments, 96 fluids have relative deviations less than 5%. Among the wet fluids ($\xi_M^* < 0$), the largest relative deviations were obtained for helium ($\Delta_{r1} = 13.86\%$ and $\Delta_{r3} = 14.38\%$) and R40 ($\Delta_{r1} = 12.65\%$ and $\Delta_{r3} = 12.83\%$). Finally, we would like to note that, even with large deviations, e.g. $\Delta_{r1} = 34.10\%$ for D6, one can obtain fairly good results, as shown in Figure 3d.

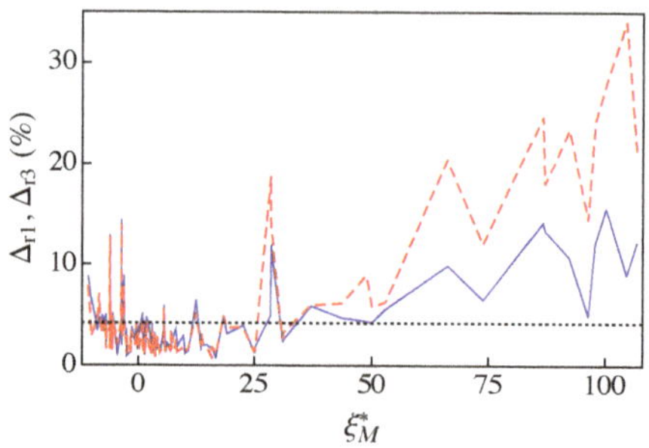

Figure 5. The percent relative deviations Δ_{r1} and Δ_{r3} vs. $\xi^*_{\rm M}$. For clarity, the deviations are plotted with lines: the dashed red line corresponds to Δ_{r1} and the solid blue line represents Δ_{r3}. The dotted line indicates a deviation level of 5%. The data for Δ_{r1}, Δ_{r3}, and $\xi^*_{\rm M}$ are listed in Table A1.

6. Summary

In a previous work [16], a semiempirical method was developed for obtaining the liquid–vapor boundary of a fluid in a $T_r - s^*$ diagram. The method assumes that, for a certain value of the quality $\bar{q} = 0.385$, the line of constant $\bar{q}$ in the $T_r - s^*$ diagram ($s^*_{\bar{q}}(T_r)$) is very close to the straight line $s^*_{\bar{q}}(T_r) \approx b(1 - T_r)$, i.e., one has a modified rectilinear diameter relation for the saturation entropies. From this assumption, the semiempirical method only requires two ingredients: an appropriate expression for the enthalpy of vaporization and an approximate estimation of the parameter b. In what refers to the enthalpy of vaporization, perhaps the simplest choice is the extended corresponding states version [20] of the Watson equation [21] considered in [16] and in the present work. Two approximations, A1 and A2, where developed for b in [16]. The more accurate is approximation A1 with $b_{\rm A1}$ given by Equation (11) and therefore depending on the acentric factor ω, the parameter $\xi^*_{\rm M}$, and the reduced temperature $T_{{\rm M}r}$. The main drawback of A1 is that, while ω is available for most fluids, the other two parameters must be calculated for each fluid. Approximation A2 is less accurate but it has the advantage that, in addition to ω, it only requires the critical molar volume v_c that can be accessed in most databases.

In the present work, we developed a new approximation A3 where the parameter $b_{\rm A3}$ is given by Equation (25) and only depends on ω and on the ideal-gas isobaric molar heat capacity of the fluid $c_p^{\rm ig*}$ at a reduced temperature $\overline{T}_{{\rm M}r} = 0.81$. We want to recall that excellent approximations for $c_p^{\rm ig*}$ are available for a large variety of fluids in most Thermophysical Properties databases. The new approximation has the virtues of approximations A1 and A2 without their problems: it has an accuracy similar to or even better than that of A1 and only needs the value of easily accessible parameters.

To conclude we would like to comment that the present work has also served to clarify the role played by the ideal-gas isobaric molar heat capacity in the shape of the liquid–vapor saturation boundary in a temperature–entropy diagram, in agreement with the results of other authors [11–15].

Author Contributions: Conceptualization, J.A.W. and S.V.; Formal analysis, J.A.W. and S.V.; and Writing—Original draft, J.A.W. and S.V.

Funding: This research was funded by Junta de Castilla y León of Spain grant number SA017P17.

Conflicts of Interest: The authors declare no conflict of interest.

Nomenclature

$-b$	Slope of the modified rectilinear diameter
Δ_r	Percent relative deviation
$\Delta_v h$	Molar enthalpy of vaporization, J/mol
v	Molar volume, m^3/kmol
c_P	Isobaric molar heat capacity, J/(mol K)
p	Pressure, MPa
q	Quality
R	Gas constant, 8.314472 J/(mol·K)
s	Molar entropy, J/(mol·K)
T	Temperature, K
Greek letters	
ω	Acentric factor
ξ	Inverse of the slope of the saturated vapor curve, J/(mol·K^2)
Superscripts	
*	Dimensionless
ex	Excess
ig	Ideal gas
r	Residual
Subscripts	
A1, A2, A3	Approximations used in this work
c	Critical
con	Condensation
ev	Evaporation
g	Saturated vapor
l	Saturated liquid
M	Point for which $\xi^*(T_r)$ presents a maximum
r	Reduced
Acronyms	
ORC	Organic Rankine cycle

Appendix A

Table A1. Critical temperature T_c, acentric factor ω, maximum slope of the entropy of the saturated vapor ξ^*_M, reduced temperature T_{Mr}, ideal gas molar isobaric heat capacity $c_p^{ig*}(\overline{T}_{Mr})$, parameters b_{A1} and b_{A3}, and percent relative deviations Δ_{r1} and Δ_{r3}. The data were obtained from RefProp 9.1 [17]. The fluids are listed in increasing value of ξ^*_M.

Fluid	T_c(K)	ω	ξ^*_M	T_{Mr}	$c_p^{ig*}(\overline{T}_{Mr})$	b_{A1}	b_{A3}	Δ_{r1}(%)	Δ_{r3}(%)
methanol	513.38	0.5625	−10.6364	0.7874	6.3594	−7.0892	−9.1848	7.74	8.78
heavy water	643.85	0.364	−9.877	0.8071	4.5129	−4.7427	−5.7266	2.86	6.06
water	647.1	0.3443	−9.846	0.8018	4.2676	−4.534	−5.3207	2.96	6.51
ammonia	405.4	0.256	−8.6109	0.8162	4.3795	−4.4225	−5.0275	4.82	3.42
hydrogen chloride	324.55	0.1288	−8.3196	0.8228	3.5036	−2.9154	−3.4131	7.06	4.78
R41	317.28	0.2004	−7.9926	0.8146	4.2702	−4.2672	−4.6467	4.02	4.02
carbon dioxide	304.13	0.2239	−7.9381	0.8219	4.1683	−4.6244	−4.6224	4.71	4.72
R32	351.26	0.2769	−7.7684	0.8198	5.0411	−5.5457	−5.9418	3.34	3.34
xenon	289.73	0.0036	−7.6918	0.8142	2.5	−1.8437	−1.7513	3.95	4.69
krypton	209.48	−0.0009	−7.6578	0.8124	2.5	−1.8225	−1.738	4.31	4.99
argon	150.69	−0.0022	−7.6561	0.8117	2.5	−1.8092	−1.7342	2.99	3.6
neon	44.49	−0.0387	−7.3103	0.8012	2.5	−1.7047	−1.6313	4.27	4.81
sulfur dioxide	430.64	0.2557	−7.2518	0.8202	5.0139	−5.7622	−5.8093	3.64	3.36
nitrous oxide	309.52	0.162	−7.0185	0.8224	4.3649	−4.6784	−4.6055	4.23	4.71
R23	299.29	0.263	−6.8638	0.8205	5.4574	−6.2521	−6.3906	3.13	2.54
carbon monoxide	132.86	0.0497	−6.8622	0.8091	3.5003	−3.328	−3.1301	3.61	5.08

Table A1. *Cont.*

Fluid	T_c(K)	ω	ξ_M^*	T_{Mr}	$c_p^{ig*}(\overline{T}_{Mr})$	b_{A1}	b_{A3}	Δ_{r1}(%)	Δ_{r3}(%)
fluorine	144.41	0.0449	−6.8234	0.8118	3.5034	−3.2888	−3.1183	1.54	1.93
nitrogen	126.19	0.0372	−6.8113	0.806	3.5004	−3.2219	−3.0899	2.61	3.62
hydrogen sulfide	373.1	0.1005	−6.8028	0.8072	4.1079	−4.1004	−4.0548	3.59	3.91
oxygen	154.58	0.0222	−6.72	0.8054	3.5014	−3.1095	−3.044	3.77	4.28
ethylene	282.35	0.0866	−6.1541	0.8156	4.4571	−4.518	−4.4365	3.39	3.97
deuterium	38.34	−0.136	−6.0597	0.7992	2.5576	−1.638	−1.4702	3.7	4.24
R40	416.3	0.243	−6.0057	0.8311	5.2349	−6.8109	−6.0242	12.65	12.83
methane	190.56	0.0114	−5.9212	0.7978	4.0063	−3.8044	−3.6343	1.57	2.11
nitrogen trifluoride	234.0	0.126	−5.8855	0.8261	5.0065	−5.3046	−5.2579	6.46	6.78
R14	227.51	0.1785	−5.7507	0.8153	5.4035	−6.1996	−5.9544	2.73	4.29
orthohydrogen	33.22	−0.218	−5.5374	0.7814	2.5	−1.1452	−1.2508	4.27	4.75
carbonyl sulfide	378.77	0.0978	−5.4797	0.8131	5.0426	−5.3573	−5.1989	1.53	2.64
ethanol	514.71	0.646	−5.4382	0.8407	10.0811	−13.1771	−14.3512	1.56	4.68
parahydrogen	32.94	−0.219	−5.4225	0.7856	2.5	−1.2215	−1.2493	4.84	5.02
hydrogen	33.15	−0.219	−5.3996	0.7873	2.5	−1.2346	−1.2493	4.94	5.04
ethane	305.32	0.0995	−4.838	0.8145	5.6208	−6.0169	−5.919	2.68	3.37
R22	369.3	0.2208	−4.8229	0.819	6.768	−7.7051	−7.8182	3.47	2.77
R161	375.25	0.216	−4.4255	0.8205	7.2552	−8.0304	−8.3988	2.05	0.98
propyne	402.38	0.204	−3.5995	0.7971	7.7056	−8.824	−8.9033	4.22	3.92
helium	5.2	−0.385	−3.5483	0.7489	2.5	−1.0479	−1.0829	13.86	14.38
R152a	386.41	0.2752	−3.4813	0.8223	8.3897	−9.8009	−10.0678	2.6	1.94
R13	302.0	0.1723	−3.3923	0.8226	7.1846	−8.4474	−8.1281	2.73	4.78
cyclopropane	398.3	0.1305	−2.9929	0.8469	7.3234	−8.2874	−8.1353	7.88	8.84
R21	451.48	0.2061	−2.9492	0.818	8.1052	−9.3764	−9.4055	4.12	3.94
propylene	364.21	0.146	−2.7352	0.8214	7.6869	−8.7418	−8.6436	3.39	4.04
R143a	345.86	0.2615	−2.4733	0.8183	9.0101	−10.6289	−10.7696	2.09	1.83
DME	400.38	0.196	−2.4475	0.8198	8.3359	−9.7307	−9.6474	3.03	3.56
trifluoroiodomethane	396.44	0.176	−2.4241	0.7992	8.3047	−9.5876	−9.526	1.1	0.85
R134a	374.21	0.3268	−1.5632	0.8172	10.3353	−12.4667	−12.7222	2.73	1.32
R12	385.12	0.1795	−1.508	0.8043	8.9565	−10.516	−10.345	1.23	2.2
propane	369.89	0.1521	−1.3906	0.8214	8.8547	−10.1711	−10.1092	3.39	3.8
R125	339.17	0.3052	−0.5398	0.8089	10.8094	−13.2288	−13.1997	2.0	2.16
acetone	508.1	0.3071	−0.3538	0.8173	11.3098	−13.3968	−13.8269	4.12	4.14
RE143a	377.92	0.289	−0.1089	0.818	11.1559	−13.3826	−13.5488	1.71	1.47
sulfur hexafluoride	318.72	0.21	−0.0175	0.8201	10.4336	−12.3555	−12.2969	3.52	3.85
R142b	410.26	0.2321	0.1386	0.8155	10.7276	−12.8389	−12.7565	2.49	2.99
R11	471.11	0.1888	0.234	0.7864	10.4086	−12.5436	−12.1762	1.53	3.64
R116	293.03	0.2566	0.4458	0.8128	11.087	−13.5038	−13.3111	2.89	4.02
cis-butene	435.75	0.202	0.9697	0.8248	11.1318	−13.219	−13.1247	4.54	5.13
R1234ze	382.51	0.313	1.0989	0.8034	12.2336	−15.0186	−14.9966	1.25	1.38
1-butene	419.29	0.192	1.3717	0.823	11.331	−13.4852	−13.3283	3.11	4.1
R1234yf	367.85	0.276	1.3815	0.8076	12.1995	−14.7444	−14.7751	1.1	0.97
R124	395.42	0.2881	1.5614	0.8091	12.4188	−15.086	−15.1036	2.17	2.07
trans-butene	428.61	0.21	1.6935	0.8185	11.7485	−14.0718	−13.9202	2.07	2.9
isobutene	418.09	0.193	1.8248	0.8193	11.7778	−13.9628	−13.8841	4.27	4.77
R141b	477.5	0.2195	2.3839	0.8112	12.3076	−14.9281	−14.6517	1.87	3.55
R115	353.1	0.248	2.9004	0.8079	12.9368	−15.8656	−15.5554	1.01	1.9
isobutane	407.81	0.184	3.044	0.8223	12.6842	−15.0477	−14.9656	3.54	4.07
R1233zd	438.75	0.305	3.0552	0.7975	13.715	−16.9118	−16.786	1.27	1.43
R123	456.83	0.2819	3.4213	0.7994	13.8353	−16.9311	−16.8226	1.04	1.31
butane	425.12	0.201	3.5432	0.8181	13.2787	−15.7971	−15.7709	3.05	3.21
R1216	358.9	0.333	3.6838	0.8095	14.2629	−17.844	−17.6026	0.8	1.36
R245fa	427.16	0.3776	4.1683	0.8193	15.2535	−18.9125	−19.0593	2.73	1.96
R236fa	398.07	0.377	4.6989	0.7995	15.7142	−19.5696	−19.6248	1.28	1.28
R114	418.83	0.2523	5.4626	0.7781	15.1045	−18.7798	−18.2512	1.94	3.41
cyclopentane	511.72	0.201	5.524	0.8268	14.8424	−17.7562	−17.7013	5.52	5.87
R236ea	412.44	0.369	5.7807	0.7922	16.5039	−20.6118	−20.5568	1.51	1.44
R245ca	447.57	0.355	5.9462	0.8074	16.4625	−20.4352	−20.4318	1.41	1.41
R227ea	374.9	0.357	6.0234	0.8071	16.5119	−20.5427	−20.5033	1.48	1.68
benzene	562.02	0.211	6.1558	0.8252	15.4544	−18.5302	−18.4997	2.11	2.28
RE245cb2	406.81	0.354	7.0324	0.796	17.3379	−21.6067	−21.5073	1.39	1.36

Table A1. *Cont.*

Fluid	T_c(K)	ω	ξ^*_M	T_{Mr}	$c_p^{ig*}(\overline{T}_{Mr})$	b_{A1}	b_{A3}	Δ_{r1}(%)	Δ_{r3}(%)
DEE	466.7	0.281	7.039	0.8045	16.8075	−20.4947	−20.4877	3.34	3.38
R218	345.02	0.3172	7.2705	0.8054	17.1169	−21.2347	−21.0463	1.59	2.63
DMC	557.0	0.346	7.3867	0.8129	16.9462	−21.7114	−20.982	2.91	2.14
R113	487.21	0.2525	7.4017	0.7761	16.6127	−20.7474	−20.1141	2.43	2.28
RE245fa2	444.88	0.387	7.9754	0.7977	18.1268	−23.0079	−22.6575	1.67	3.46
neopentane	433.74	0.1961	8.1324	0.8216	16.828	−20.3068	−20.1319	2.49	3.58
isopentane	460.35	0.2274	8.3612	0.8179	17.3273	−20.9858	−20.8834	1.81	2.43
pentane	469.7	0.251	8.4366	0.816	17.5954	−21.4005	−21.3204	1.33	1.81
RC318	388.38	0.3553	9.8443	0.7961	19.431	−24.4359	−24.0982	1.76	2.96
R365mfc	460.0	0.377	10.1649	0.8151	19.9073	−24.9196	−24.8015	1.28	1.17
toluene	591.75	0.2657	10.8934	0.8121	19.8028	−24.0839	−24.1134	1.61	1.49
cyclohexane	553.6	0.2096	12.5657	0.8378	20.4864	−24.9198	−24.706	5.14	6.45
hexane	507.82	0.299	13.6814	0.8127	22.3454	−27.3396	−27.4114	1.87	1.46
isohexane	497.7	0.2797	13.7711	0.8125	22.2099	−27.1569	−27.1511	3.02	3.05
RE347mcc	437.7	0.403	14.2038	0.7903	23.5211	−29.5492	−29.405	2.02	1.97
perfluorobutane	386.33	0.371	14.8856	0.7876	23.6981	−29.8004	−29.4493	1.76	2.0
m-xylene	616.89	0.326	16.0966	0.8077	24.2987	−30.1694	−29.9569	0.7	1.58
p-xylene	616.17	0.324	16.0973	0.8127	24.3608	−30.11	−30.0235	1.82	1.82
ethylbenzene	617.12	0.305	16.926	0.8164	24.8511	−30.6508	−30.5343	1.03	0.68
o-xylene	630.26	0.312	17.4037	0.8105	25.3699	−31.2593	−31.2093	2.15	2.2
methylcyclohexane	572.2	0.234	18.5937	0.8222	25.7539	−31.2969	−31.316	4.93	4.82
heptane	540.13	0.349	19.2387	0.8059	27.2864	−33.6527	−33.7634	3.71	3.11
perfluoropentane	420.56	0.423	22.8082	0.7793	30.5375	−38.5956	−38.1795	3.7	3.92
octane	569.32	0.395	25.1117	0.8011	32.3926	−40.2261	−40.3133	1.37	1.19
isooctane	544.0	0.303	25.2627	0.8145	31.6407	−38.968	−38.9066	1.38	1.72
novec649	441.81	0.471	28.6415	0.7003	34.7076	−47.1405	−43.6078	18.68	4.96
MM	518.7	0.418	28.6999	0.7645	35.4931	−44.6629	−44.2692	13.84	11.87
nonane	594.55	0.4433	31.1834	0.7963	37.5659	−47.0458	−46.9732	2.59	2.29
propylcyclohhexane	630.8	0.326	31.4514	0.8077	37.0298	−45.5243	−45.6744	3.26	2.55
decane	617.7	0.4884	37.192	0.7864	42.8909	−53.8373	−53.8158	5.94	5.84
undecane	638.8	0.539	43.7309	0.7863	48.299	−61.1241	−60.8093	6.12	4.7
MDM	564.09	0.529	48.9411	0.7539	51.8895	−66.8028	−65.1781	8.89	4.38
dodecane	658.1	0.574	50.0562	0.7743	53.7356	−68.1724	−67.75	5.8	4.17
D4	586.49	0.592	52.9243	0.7727	56.3406	−71.3409	−71.0868	6.19	5.58
MD2M	599.4	0.668	66.2116	0.7353	66.5042	−86.7477	−84.1675	20.4	9.87
D5	619.23	0.658	73.8466	0.7473	73.2238	−93.8744	−92.391	11.98	6.48
methyl palmitate	755.0	0.91	86.6279	0.7388	84.2017	−110.949	−107.9614	24.68	14.12
methyl linolenate	772.0	1.14	87.0511	0.7625	86.3193	−114.2828	−112.7746	17.94	13.3
MD3M	628.36	0.722	92.3726	0.7165	87.7973	−114.4347	−110.8565	23.44	10.69
methyl linoleate	799.0	0.805	96.4907	0.7361	91.2891	−119.2024	−115.8209	14.41	4.83
methyl oleate	782.0	0.906	97.9157	0.7395	92.9609	−122.1459	−118.7401	24.02	12.09
methyl stearate	775.0	1.02	100.2849	0.7366	95.3592	−126.492	−122.7452	27.94	15.56
D6	645.78	0.736	104.5704	0.6897	95.7557	−128.0031	−120.7889	34.1	8.95
MD4M	653.2	0.825	106.7797	0.7384	100.342	−129.7364	−127.1614	21.34	12.13

References

1. Saleh, B.; Koglbauer, G.; Wendland, M.; Fischer, J. Working fluids for low-temperature organic Rankine cycles. *Energy* **2007**, *32*, 1210–1221, doi:10.1016/j.energy.2006.07.001. [CrossRef]
2. Quoilin, S.; Declaye, S.; Tchanche, B.F.; Lemort, V. Thermo-economic optimization of waste heat recovery Organic Rankine Cycles. *Appl. Therm. Eng.* **2011**, *31*, 2885–2893, doi:10.1016/j.applthermaleng.2011.05.014. [CrossRef]
3. Wang, E.; Zhang, H.; Fan, B.; Ouyang, M.; Zhao, Y.; Mu, Q. Study of working fluid selection of organic Rankine cycle (ORC) for engine waste heat recovery. *Energy* **2011**, *36*, 3406–3418.10.1016/j.energy.2011.03.041. [CrossRef]
4. SprouseIII, C.; Depcik, C. Review of organic Rankine cycles for internal combustion engine exhaust waste heat recovery. *Appl. Therm. Eng.* **2013**, *51*, 711–722, doi:10.1016/j.applthermaleng.2012.10.017. [CrossRef]
5. Bao, J.; Zhao, L. A review of working fluid and expander selections for organic Rankine cycle. *Renew. Sust. Energ. Rev.* **2013**, *24*, 325–342, doi:10.1016/j.rser.2013.03.040. [CrossRef]

6. Hærvig, J.; Sørensen, K.; Condra, T. Guidelines for optimal selection of working fluid for an organic Rankine cycle in relation to waste heat recovery. *Energy* **2016**, *96*, 592–602, doi:10.1016/j.energy.2015.12.098. [CrossRef]
7. Liu, B.T.; Chien, K.H.; Wang, C.C. Effect of working fluids on organic Rankine cycle for waste heat recovery. *Energy* **2004**, *29*, 1207–1217, doi:10.1016/j.energy.2004.01.004. [CrossRef]
8. Györke, G.; Deiters, U.K.; Groniewsky, A.; Lassu, I.; Imre, A.R. Novel classification of pure working fluids for Organic Rankine Cycle. *Energy* **2018**, *145*, 288–300, doi:10.1016/j.energy.2017.12.135. [CrossRef]
9. Györke, G.; Groniewsky, A.; Imre, A.R. A Simple Method of Finding New Dry and Isentropic Working Fluids for Organic Rankine Cycle. *Energies* **2019**, *12*, 480. [CrossRef]
10. Imre, A.R.; Kustán, R.; Groniewsky, A. Thermodynamic Selection of the Optimal Working Fluid for Organic Rankine Cycles. *Energies* **2019**, *12*, 2028. [CrossRef]
11. Morrison, G. The shape of the temperature-entropy saturation boundary. *Int. J. Refrig* **1994**, *17*, 494–504, doi:10.1016/0140-7007(94)90011-6. [CrossRef]
12. Garrido, J.M.; Quinteros-Lama, H.; Mejía, A.; Wisniak, J.; Segura, H. A rigorous approach for predicting the slope and curvature of the temperature-entropy saturation boundary of pure fluids. *Energy* **2012**, *45*, 888–899.10.1016/j.energy.2012.06.073. [CrossRef]
13. Albornoz, J.; Mejía, A.; Quinteros-Lama, H.; Garrido, J.M. A rigorous and accurate approach for predicting the wet-to-dry transition for working mixtures in organic Rankine cycles. *Energy* **2018**, *156*, 509–519, doi:10.1016/j.energy.2018.05.074. [CrossRef]
14. Groniewsky, A.; Györke, G.; Imre, A.R. Description of wet-to-dry transition in model ORC working fluids. *Appl. Therm. Eng.* **2017**, *125*, 963–971, doi:10.1016/j.applthermaleng.2017.07.074. [CrossRef]
15. Groniewsky, A.; Imre, A.R. Prediction of the ORC Working Fluid's Temperature-Entropy Saturation Boundary Using Redlich-Kwong Equation of State. *Entropy* **2018**, *20*, 93, doi:10.3390/e20020093. [CrossRef]
16. White, J.A.; Velasco, S. A Simple Semiempirical Method for Predicting the Temperature-Entropy Saturation Curve of Pure Fluids. *Ind. Eng. Chem. Res.* **2019**, *58*, 1038–1043. [CrossRef]
17. Lemmon, E.W.; Huber, M.L.; McLinden, M.O. *NIST Standard Reference Database 23: Reference Fluid Thermodynamic and Transport Properties-REFPROP, Version 9.1. Standard Reference Data Program*; National Institute of Standards and Technology: Gaithersburg, MD, USA, 2013.
18. Bell, I.H.; Wronski, J.; Quoilin, S.; Lemort, V. Pure and Pseudo-pure Fluid Thermophysical Property Evaluation and the Open-Source Thermophysical Property Library CoolProp. *Ind. Eng. Chem. Res.* **2014**, *53*, 2498–2508. [CrossRef]
19. Mulero, A.; Cachadina, I.; Parra, M.I. Comparison of corresponding-states-based correlations for the prediction of the vaporization enthalpy of fluids. *Ind. Eng. Chem. Res.* **2008**, *47*, 7903–7916. [CrossRef]
20. Velasco, S.; Santos, M.J.; White, J.A. Extended corresponding states expressions for the changes in enthalpy, compressibility factor and constant-volume heat capacity at vaporization. *J. Chem. Thermodyn.* **2015**, *85*, 68–76, doi:10.1016/j.jct.2015.01.011. [CrossRef]
21. Watson, K. Prediction of critical temperatures and heats of vaporization. *Ind. Eng. Chem.* **1931**, *23*, 360–364. [CrossRef]
22. Cailletet, L.; Mathias, E. Recherches sur les densités des gaz liquéfiés et de leurs vapeurs saturées. *J. Phys. Theor. Appl.* **1886**, *5*, 549–564. [CrossRef]
23. White, J.; Velasco, S. Characterizing wet and dry fluids in temperature-entropy diagrams. *Energy* **2018**, *154*, 269–276, doi:10.1016/j.energy.2018.04.105. [CrossRef]
24. Aly, F.A.; Lee, L.L. Self-consistent equations for calculating the ideal gas heat capacity, enthalpy, and entropy. *Fluid Phase Equilib.* **1981**, *6*, 169–179, doi:10.1016/0378-3812(81)85002-9. [CrossRef]
25. Rowley, R.; Wilding, W.; Oscarson, J.; Yang, Y.; Zundel, N.; Daubert, T.; Danner, R. DIPPR data compilation of pure chemical properties. In *DIPPR Data Compilation of Pure Chemical Properties*; Design Institute for Physical Properties: New York, NY, USA, 2006.

energies

MDPI

Article

Working Fluid Selection for Organic Rankine Cycle Using Single-Screw Expander

Xinxin Zhang [1,2,*], Yin Zhang [1,2], Min Cao [1,2], Jingfu Wang [1,2], Yuting Wu [1,2] and Chongfang Ma [1,2]

1 MOE Key Laboratory of Enhanced Heat Transfer and Energy Conservation, College of Environmental and Energy Engineering, Beijing University of Technology, Beijing 100124, China

2 Beijing Key Laboratory of Heat Transfer and Energy Conversion, College of Environmental and Energy Engineering, Beijing University of Technology, Beijing 100124, China

* Correspondence: xinxinzhang@bjut.edu.cn; Tel.: +86-10-6739-1985

Received: 6 July 2019; Accepted: 15 August 2019; Published: 20 August 2019

Abstract: The organic Rankine cycle (ORC) is a popular technology used in waste heat recovery and medium-low-temperature heat utilization. Working fluid plays a very important role in ORC. The selection of working fluid can greatly affect the efficiency, the operation condition, the impact on the environment, and the economic feasibility of ORC. The expander is a key device in ORC. As a novel expander, single-screw expanders have been becoming a research focus in the above two areas because of their many good characteristics. One of the advantages of single-screw configurations is that they can conduct a vapor–liquid two-phase expansion. Therefore, in order to give full play to this advantage, a working fluid selection for ORC using a single-screw expander was conducted in this paper. Three indicators, namely, net work output, thermal efficiency, and heat exchange load of condenser, were used to analyze the performance of an ORC system. Through calculation and analysis, it can be seen that an ORC system that uses a single-screw expander and undergoes a vapor–liquid two-phase expansion is able to obtain a higher thermal efficiency, higher net work output, and a smaller heat exchange load of the condenser. Regardless of whether isentropic efficiency of the expander is considered or not, cis-butene may be the best candidate for working in subcritical cycles. HFO working fluids are more suitable for working in transcritical cycles, and HFO-1234ze(E) may be the best.

Keywords: single-screw expander; vapor–liquid two-phase expansion; thermal efficiency; net work output; heat exchange load of condenser; cis-butene; HFO-1234ze(E)

1. Introduction

Renewable and sustainable energy utilization and recovery of low-grade waste heat are two measures for alleviating the global energy crisis and for solving the environmental problems caused by the consumption of traditional energy sources. The organic Rankine cycle (ORC) has been widely studied and adopted in the two fields mentioned above. Working fluid selection plays a decisive role in an ORC, because it can greatly affect thermodynamic performance, working conditions, impact on the environment, and economic feasibility [1,2]. Basically, organic working fluid can be classified into three categories, on the basis of the slope of the saturated vapor curve in a *T-s* diagram. These are: dry fluid, with a positive slope; wet fluid, with a negative slope; and isentropic fluid, with a vertical slope [3]. From the viewpoint of thermodynamic performance, dry and isentropic working fluids are more appropriate for ORC systems, because after isentropic expansion, they are in superheated and saturated states, respectively. Therefore, damage to the expander caused by wet vapor can be avoided. As a result of environmental concerns, the CFCs and HCFCs that dominated organic working fluids from 1931 until the early 1990s have been and are being phased out, respectively. Nowadays, HFO working fluids are drawing increasing attention. R1234yf and R1234ze are two typical HFO working fluids that have recently been studied and are being used frequently [4–7]. As far as the economic performance of a practical ORC system is concerned, the investment required by the heat exchanger accounts for the largest proportion, followed by that for the expander, and the working fluid accounts for the third [8]. The first two investments are significantly influenced by the characteristics of the working fluid.

The expander is a critical device in an ORC system, because it determines the thermodynamic performance of the ORC system, and its cost ranks second in the total system investment [8]. Expanders, can generally be categorized into two types: the turbo type and the positive-displacement type. Turbo expanders are normally suitable for large-scale ORC systems [9,10]. However, they might not be favorable for small-scale ORC units [11]. Positive-displacement expanders, such as rolling piston expanders, scroll expanders, and single-screw expanders, are good substitutes for turbo machines due to their relatively high efficiency, high pressure ratio, low rotational speed, and tolerance of two-phase fluids [9]. Among the above three types, scroll expanders constitute a research hotspot. However, their power is normally lower than 3.5 kW. Lemort et al. tested four volumetric expanders (scroll, screw, piston and roots) at <5 kW in a small-scale organic Rankine cycle system with R245fa as the working fluid. They found that the scroll expander showed the highest isentropic efficiency [12]. In contrast, single-screw expanders seem more promising when the output power is higher than 3.5 kW [11]. Furthermore, screw expanders are more technically mature than scroll and piston expanders [13]. The single-screw configuration was first invented in the 1950s, and has mainly been employed in compressors [14,15]. Nowadays, this configuration is widely used in the design and manufacture of expanders. Compared with rolling piston expanders and scroll expanders, single-screw expanders have many advantages, such as balanced load of the screw, long service life, high volume efficiency, good performance at partial load, low leakage, low noise, low vibration, and simple configuration [11].

All of these advantages have attracted many researchers to carry out relevant study on the design and application of single-screw expanders. The group of Lemort and Quoilin modified a standard compressor into an 11 kW single-screw expander and used it to conduct an experimental study. The maximum expander isentropic efficiency and the generated power were, respectively, 64.78% and 7.8 kW [16]. Ziviani et al. carried out a comprehensive review of the geometry modeling of single-screw machines. In their review, the main geometric parameters and constraints were discussed, and the limitations of the existing methodologies were highlighted. An 11 kW single-screw expander was considered in carrying out the calculations [17]. Giuffrida improved the semi-empirical modeling of single-screw expanders for small organic Rankine cycles, and the application of the proposed modeling to a single-screw expander resulted in mean absolute percentage errors of 0.69%, 1.77% and 0.33% with respect to mass flow rate, electric power output and exhaust fluid temperature, respectively [18]. Wang et al. established a theoretical model for analyzing the internal leakage of single-screw expanders [19], and studied the influence of clearance height on the performance of single-screw expanders in small-scale organic Rankine cycles [20]. They found that an appropriate fitting gap adjustment was a necessary measure for improving single-screw expander performance. The volumetric efficiency decreased by 4.42–7.7%, and the power output increased by 0.02–0.11 kW, when the fitting clearance height was between 0.03 mm and 0.09 mm. They also found that clearance height had a great influence on the expander performance. The volumetric and isentropic efficiency decreased slowly at first, and then more quickly with increasing clearance height at a given rotation speed.

Experimental studies on single-screw expanders initially used compressed air as the working fluid. Lu et al. established a compressed air refrigeration system using a single-screw expander with 175 mm diameter [21]. Higher than 65% adiabatic efficiency for single-screw expander was achieved and a more than 70 °C temperature drop could be obtained in the experiment. He et al. conducted an experiment to study the influence of intake pressure on the performance of a 175-mm-diameter single-screw expander working with compressed air [22]. They found that the highest overall efficiency could reach 55% or so, that the greatest torque reached nearly 100 N·m, that the highest power output amounted to about 22 kW, and that the lowest gas consumption was about 60 kg/kWh.

Subsequently, some work was done on single-screw expanders employed in ORC systems for the purposes of waste heat recovery. Zhang et al. developed a single-screw expander with a 155-mm-diameter screw and established an ORC system with R123 as the working fluid for waste heat recovery from the exhaust of a diesel engine [23]. They achieved a maximum power output of 10.38 kW. The highest ORC efficiency and overall system efficiency were 6.48% and 43.8%, respectively, which were achieved at a diesel engine output of 250 kW. Yang et al. conducted similar work using R245fa as the working fluid [24]. They presented a vehicle diesel engine–ORC combined system. They found that when the diesel engine speed was 2200 r/min and the diesel engine torque was 1200 N·m, the power output of the combined system reached its maximum value, at approximately 308.6 kW, which is 28.6 kW higher than the power output of the diesel engine.

Based on the reviewed papers, it can be seen that thermodynamic cycle analysis has previously generally been based solely on the use of single-screw expanders as an alternative to other types of expanders. Working fluid selection has also typically been based on the traditional organic Rankine cycle model. Wajs et al. presented a prototype for a domestic ORC micropower plant that used a gas boiler as an autonomous source of heat. They found that a domestic gas boiler was able to provide the saturated/superheated ethanol vapor and steam necessary to act as working fluids for ORC and RC systems, respectively [25]. Dariusz Mikielewicz and Jaroslaw Mikielewicz proposed a thermodynamic criterion for working fluid selection, for both the subcritical and the supercritical organic Rankine cycle. They chose R123 and R141b for small-scale domestic CHP applications [26]. They also suggested an Organic Flash Cycle (OFC) that would potentially be able to improve the utilization efficiency of the heat source. In their study, the results showed that the single flash OFC achieves better efficiencies than the optimized basic ORC [27]. R245fa and SES36 were selected by the group of Lemort and Quoilin in a test rig equipped with a single-screw expander that had been modified from a standard compressor with a shaft power of 11 kW. The reason for the selection of this working fluid selection was simple: null ODP (ozone depletion potential). This is a well-known factor in the power industry, especially among ORC manufacturers [28]. Tolerance of vapor–liquid two-phase expansion, which is an important characteristic of single-screw expanders, has not been considered or utilized in previous analysis and research. The selection of working fluids that match the characteristics of vapor–liquid two-phase expansion in single-screw expanders has not been carried out, although in a review paper, White et al. compared the power output predicted by two different models of n-pentane, n-hexane, and isopentane within partially evaporated and a superheated cycles [29]. The selection of the expander and the working fluid are strongly interconnected. Therefore, in this paper, two ORC models—a subcritical cycle model and a transcritical cycle model—that are able to give full play to the advantages of vapor–liquid two-phase expansion are presented. Moreover, working fluid selection is conducted for ORC using a single-screw expander on the basis of three indicators, namely, net work output, thermal efficiency, and heat exchange load of condenser.

2. Ideal Subcritical Cycle Model

Liu et al. stated that dry or isentropic working fluids were the most appropriate for ORC systems. This is because these two types of working fluid are superheated following isentropic expansion; thereby, any damage to the turbine blades caused by liquid droplets can be avoided [3]. They also concluded that a superheated apparatus was not required. This conclusion was with reference to a turbo-type expander. However, for single-screw expanders, which are able to tolerate vapor–liquid two-phase expansion, different conclusions may be drawn. Figure 1 depicts the configuration of a single-screw expander. Its working processes are depicted by Figure 2.

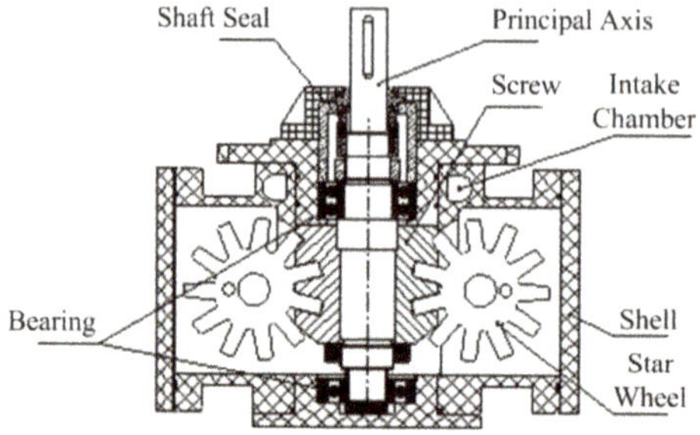

Figure 1. Configuration of a single-screw expander.

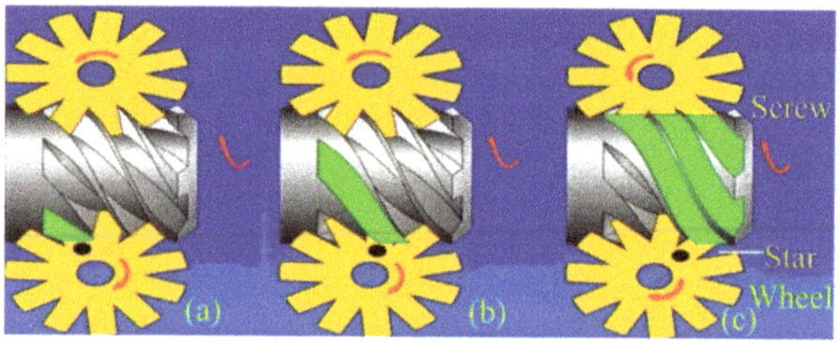

Figure 2. Working processes of single-screw expanders: (**a**) suction, (**b**) expansion, and (**c**) discharge.

It should be noted, here, that although single-screw expanders are able to carry out vapor–liquid two-phase expansion, wet working fluids have still not been adopted in ORC systems with single-screw expanders as a result of three considerations. First, there is no consensus regarding the minimum dryness of the working fluid used with single-screw expanders. Second, the expansibility of vapor-phase working fluids is much better than that of liquid-phase working fluids. Last, low- to medium-grade thermal energy is "precious", and it cannot be wasted on overheating wet working fluids. Therefore, in the following calculation and analysis, only dry and isentropic working fluids are considered and included.

2.1. Working Conditions of Expander Inlet are Known

2.1.1. Thermodynamic Setting and Description

In a *T-s* diagram, a significant difference between dry (or isentropic) fluid and wet fluid is the existence of a point at which the entropy value reaches the maximum on the saturated vapor curve ranging from the normal boiling point to the critical point. This point is located near the critical point, and is defined as the turning point [30]; other details about the role of this point can be found elsewhere [31]. The turning point temperature is the limit of subcritical ORC when adopting a traditional expander [30]. However, when using a single-screw expander, a subcritical ORC process that is capable of taking advantage of the vapor–liquid two-phase expansion of single-screw expander can be established, as is depicted in Figure 3.

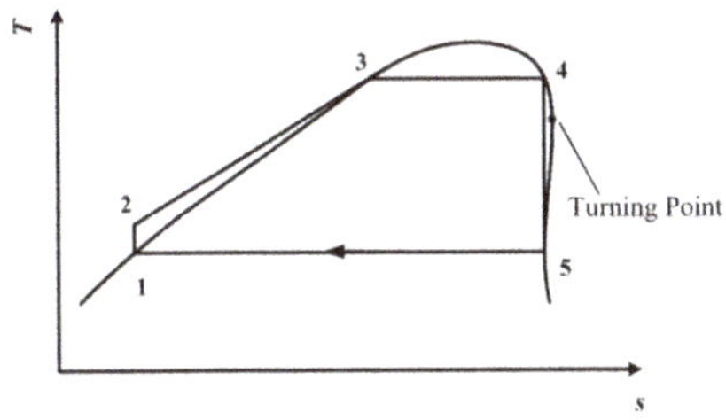

Figure 3. Subcritical ORC with a single-screw expander when the working conditions of the expander inlet are known.

In Figure 3, net work output is calculated by

$$w_{\text{net}} = (h_4 - h_5) - (h_2 - h_1), \tag{1}$$

thermal efficiency is calculated by

$$\eta = \frac{w_{\text{net}}}{q_e} = \frac{(h_4 - h_5) - (h_2 - h_1)}{h_4 - h_2}, \tag{2}$$

and the heat exchange load of the condenser is calculated by

$$q_c = h_5 - h_1, \tag{3}$$

In the above equations, h is enthalpy, w is work, q is heat exchange, η is thermal efficiency, c stands for condenser, e stands for evaporator, and the numbers are the state points in the figure. In the following equations, h, w, q, η, and the numbers all have the same meanings.

We hope that the whole cycle depicted in Figure 3 will still operate in the subcritical region. Meanwhile, vapor–liquid two-phase expansion should be ensured. Therefore, both state point 4 and state point 5 should be on the saturated vapor curve. The turning point should be located between points 4 and 5. To be more specific, $0.9T_c$ (critical temperature) represents the extreme temperature of the subcritical region. The temperature at state point 4 is set to be this extreme temperature. That is to say, $0.9T_c$ (critical temperature) should be higher than the temperature at the turning point.

From Figure 3, it can be seen that the process 4–5 is a two-phase isentropic expansion. Both state point 4 and state point 5 are on the saturated vapor curve. On the contrary, traditional subcritical ORC with dry/isentropic working fluid is depicted in Figure 4. In Figure 4, state point 5, the end of the isentropic expansion, is in a superheated state. To achieve the best thermodynamic performance in the traditional working mode, the temperature at point 4, which is also the expander inlet temperature, is set to be the same as the temperature at the turning point. State point 4 is still on the saturated vapor curve. The same condensation temperature is used for both Figures 3 and 4.

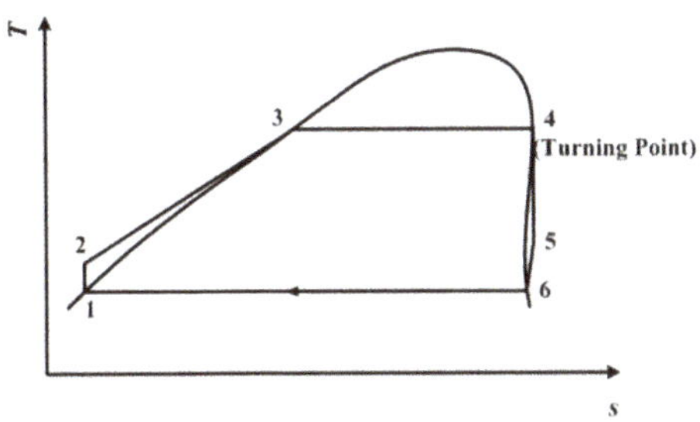

Figure 4. Traditional subcritical ORC with dry/isentropic working fluid.

In Figure 4, net work output is calculated by

$$w_{\text{net}} = (h_4 - h_5) - (h_2 - h_1), \tag{4}$$

thermal efficiency is calculated by

$$\eta = \frac{w_{\text{net}}}{q_e} = \frac{(h_4 - h_5) - (h_2 - h_1)}{h_4 - h_2}, \tag{5}$$

and the heat exchange load of the condenser is calculated by

$$q_c = h_5 - h_1, \tag{6}$$

Here, it should be observed that the maximum operating temperature of the single-screw expander should not exceed 130 °C (400 K), due to the restriction of the sealing material, lubricating oil, and starwheel material. Therefore, the temperature at state point 4 cannot exceed 130 °C (400 K) in either the two-phase expansion process or the traditional working process. Therefore, single-screw expanders are especially suitable for biomass combined with heat and power, the temperature of which varies between 150 and 320 °C, and heat recovery for mechanical equipment and industrial processes, the temperatures of which vary between 215 °C and 315 °C [32].

2.1.2. Results and Discussion

Taking into account the above settings and limitations, all the working fluids included in REFPROP 9.1 [33] were screened. A total of 73 dry and isentropic working fluids were found. Among these 73 working fluids, there were 6 working fluids that were able to meet all of the thermodynamic requirements described above. These were cis-butene, R11, R116, R1234yf, R1234ze(E), and R142b. The main thermodynamic, safety, and environmental properties of the above 6 working fluids are listed in Table 1. Based on the data listed in Table 1, it can be seen that R11, R116, and R142b are not suitable for use in ORC systems with single-screw expanders due to their high GWP value. Among the remaining three working fluids, R1234yf and R1234ze(E) are popular HFO working fluids that have been widely studied. As for cis-butene, it belongs to the HC working fluids and belongs to olefins, chemically. Its safety group could not be found in ANSI/ASHRAE Standard 34. However, some statements about it with respect to risk and safety were found. Hazard and precautionary statements were also found. All these statements are summarized as follows [34]:

- Extremely flammable gas.
- Autoignition Temperature: 615 °F (323.89 °C)
- Protect from sunlight.
- Keep container in a well-ventilated place.
- Keep away from sources of ignition—No smoking.
- Take precautionary measures against static discharges.

Table 1. Main thermodynamic, safety, and environmental properties of 6 working fluids.

Working Fluid	Critical Temperature (T_c)/K	The extreme Temperature of Subcritical Region ($0.9T_C$)/K	Turning Point Temperature/K	Safety Group	Global Warming Potential (GWP), 100 year
cis-butene(cis-2-butene)	435.75	392.18	390	N/A	~20 [35]
R11	471.11	424.00	395	A1 [36]	4750 [37]
R116	293.03	263.73	255	A1 [36]	5700 [38]
R1234yf	367.85	331.07	330	A2L [1] [39–41]	<1 [42]/4 [43]
R1234ze(E)	382.51	344.26	340	A2L [2] [39–41]	<1 [42]/6 [44]
R142b	410.26	369.24	350	A2 [36]	2310 [37]

[1,2] A2L is a lower flammability refrigerant with a maximum burning velocity of ≤10 cm/s.

Low-grade waste heat can be divided into two categories: open type and closed type [45,46]. For the open type, the inlet temperature and mass flow rate are known, and the working mass of the heat source is directly discharged after being used. For the closed type, the heat release is specific, and the working mass of the heat source is usually recycled after releasing heat. Therefore, the standards used to measure the waste heat recovery of these two types of heat source are different [45]. The maximum net power output is used as the criterion for the open type, while the maximum thermal efficiency is used for the closed type. Therefore, net power output and thermal efficiency are adopted as the first two indicators for evaluating the performance of the vapor–liquid two-phase expansion working mode and the traditional working mode of ORC. The third evaluation indicator is the heat exchange load of the condenser, because this is critical for calculating the cost of the condenser, which will greatly influence the cost and economic performance of entire ORC system [8].

Considering the thermodynamic requirements with respect to Figure 3, the entropy value of point 4 (s_4) is determined by its temperature ($T_4 = 0.9T_C$). Therefore, the temperature of point 5 (T_5) can be determined on the basis of the entropy value of point 5 ($s_5 = s_4$).

Table 2 lists the temperatures at point 4 and 5, along with the thermodynamic performance indicated by the above three indicators for the remaining three working fluids in the model of the subcritical cycle. From Table 2, it can be seen that in vapor–liquid two-phase expansion mode, the condensation temperature, which is the same as the expander outlet temperature of R1234yf, is too low, and is beyond the usual condensation temperature range. Therefore, its thermodynamic performance is not calculated and discussed. As for R1234ze(E), the shape of its saturated vapor curve, depicted in Figure 5, mean that it has two intersections (point 5) with the isentropic line. One is at 331 K, the other is at 237.8 K, which is also too low, and is beyond the usual condensation temperature range. Therefore, 331 K is designated as the condensation temperature for the purposes of further discussion. Based on the data listed in Table 2, it can be seen that cis-butene has a suitable condensation temperature, resulting in an appropriate temperature difference between the expander inlet and the outlet, which is the same as the condensation temperature. However, the temperature difference between the expander inlet and the outlet of R1234ze(E) is very small. This results in a low net power output and a low thermal efficiency, although it also leads to a small heat exchange load for the condenser. When comparing the vapor–liquid two-phase expansion mode with the traditional mode, it can be seen that the vapor–liquid two-phase expansion mode has a better thermodynamic performance, specifically, a higher thermal efficiency, a higher net power output, and a smaller heat exchange load for the condenser. However, the difference in thermodynamic performance between the above two modes is very small, due to the small difference in temperature at point 4 between the two working modes.

Table 2. Thermodynamic performance of the remaining three working fluids in the subcritical cycle model when the working conditions at the expander inlet are known.

Working Fluid	Working Mode	Expander Inlet Temperature T_4/K	Expander Outlet Temperature T_5/K	Net Power Output/kJ·kg^{-1}	Thermal Efficiency/%	Heat Exchange Load of Condenser/kJ·kg^{-1}
cis-butene (cis-2-butene)	Vapor–liquid two-phase expansion mode	392.18	283.1	113.88	21.82	408.06
	Traditional mode	390	283.12	112.40	21.59	408.10
R1234yf	Vapor–liquid two-phase expansion mode	331.07	223.5 [1]	N/A	N/A	N/A
	Traditional mode	330	223.51 [2]	N/A	N/A	N/A
R1234ze(E)	Vapor–liquid two-phase expansion mode	344.26	331(237.8 [3])	5.11	3.58	137.78
	Traditional mode	340	331.08	3.56	2.52	137.87

[1,2,3] The temperature is too low and beyond the usual condensation temperature range.

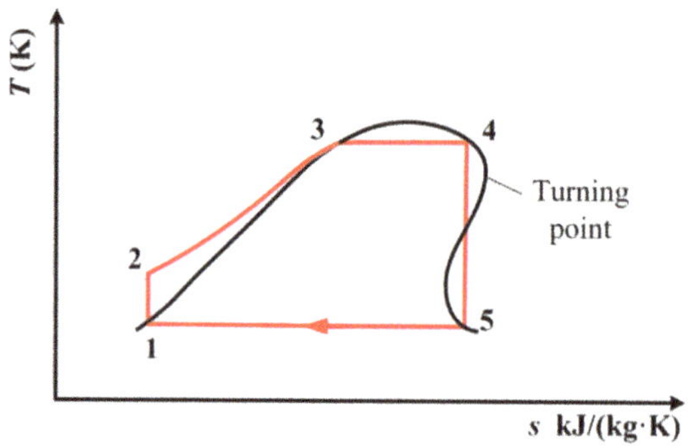

Figure 5. Shape of the saturated vapor curve of R1234yf, resulting in two intersections with the isentropic line.

In summary, cis-butene can be used as a suitable working fluid in the subcritical cycle model when the working conditions at the expander inlet are known. It is suitable for both open and closed types of heat source. However, the cost of the heat exchanger for cis-butene will be higher than that for R1234yf and R1234ze(E).

2.2. Condensation Temperature is Known

2.2.1. Thermodynamic Setting and Description

The analysis provided in the previous subsection was based on the operating conditions of expander inlet being known, which will act as a determinant on condensation temperature. The results show that the difference in thermodynamic performance between the two working modes is very small. Therefore, in this subsection, we will first attempt to determine and set the condensation temperature, and subsequently conduct the thermodynamic performance analysis.

According to [47], 320 K and 290 K are the recommended condensation temperatures for working fluids with high and low critical temperatures, respectively. These two condensation temperatures can be achieved using air cooling and water cooling. In order to perform a detailed analysis, the thermodynamic performance of the above three working fluids are calculated with condensation temperature varying from 290 K to 320 K. The whole cycle process for cis-butene is depicted in Figure 6. The process for R1234yf and R1234ze(E) are depicted in Figure 7. In Figures 6 and 7, the temperature at point 4(T_4) is still set as the extreme temperature of the subcritical ORC, which is $0.9T_c$ (critical temperature). Line 4′–5 is an isentropic line. Point 5 is on the saturated vapor curve. The temperature at point 5 (T_5) is the same as the condensation temperature, which is varied between 290 K and 320 K. From Figures 6 and 7, it can be seen that point 4′ is not in a saturated state, and has a vapor quality.

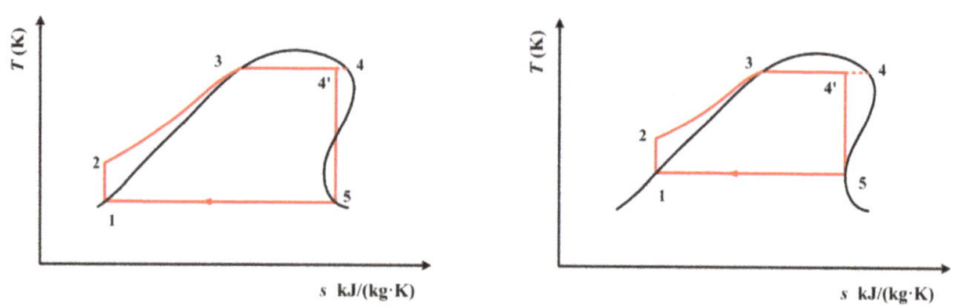

Figure 6. Subcritical ORC for cis-butene with a single-screw expander when the condensation temperature is known: (**a**) 290 K, (**b**) 320 K.

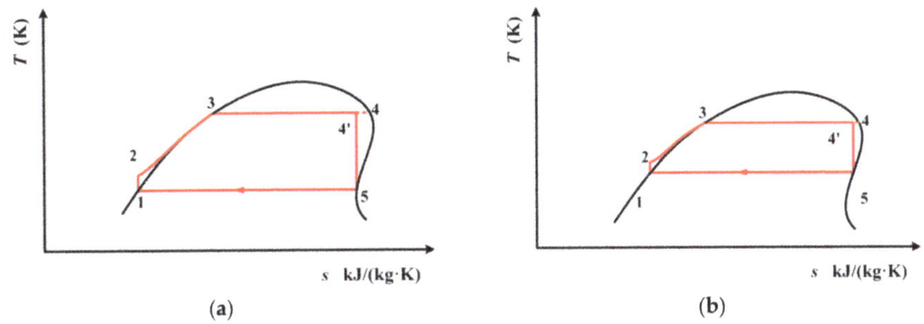

Figure 7. Subcritical ORC for R1234yf and R1234ze(E) with a single-screw expander when the condensation temperature is known: (**a**) 290 K, (**b**) 320 K.

In Figures 6 and 7, the net work output is calculated by

$$w_{\text{net}} = (h_{4'} - h_5) - (h_2 - h_1), \tag{7}$$

thermal efficiency is calculated by

$$\eta = \frac{w_{\text{net}}}{q_e} = \frac{(h_{4'} - h_5) - (h_2 - h_1)}{h_{4'} - h_2}, \tag{8}$$

the heat exchange load of the condenser is calculated by

$$q_c = h_5 - h_1, \tag{9}$$

and the vapor quality of the working fluid is calculated by

$$x = \frac{s_{4'} - s_3}{s_4 - s_3}, \tag{10}$$

It should be noted, here, that for cis-butene, 320 K only corresponds to the local minimum point of entropy on its saturated vapor curve, as shown in Figure 6b. Meanwhile, for R1234yf and R1234ze(E), 290 K, which is their lowest condensation temperature, is higher than the temperature which corresponds to the local minimum entropy.

The subcritical ORC depicted in Figure 4 is still used for the purpose of comparison. The temperature at point 4, which is the same as the temperature of the expander inlet, is still set to be same as the temperature at the turning point. State point 4 is still on the saturated vapor curve. Point 6 is on the saturated vapor curve, and its temperature (T_6) is the same as the condensation temperature, which is varied between 290 K and 320 K.

2.2.2. Results and Discussion

Table 3 lists the net power output, thermal efficiency, and heat exchange load of the condenser for cis-butene, R1234yf, and R1234ze(E) when the condensation temperature is known.

Table 3. Thermodynamic performance of the three working fluids in the subcritical cycle model when the condensation temperature is known.

Working Fluid	Working Mode	Condensation Temperature T_5/K	Condensation Temperature T_6/K	Net Power Output/kJ·kg^{-1}	Thermal Efficiency/%	Heat Exchange Load of Condenser/kJ·kg^{-1}	Vapor Quality
cis-butene (cis-2-butene)	Vapor–liquid two-phase expansion mode	284.7	/	111.42	21.51	406.46	0.9978
		290		103.48	20.51	401.09	0.9913
		295		96.33	19.57	395.91	0.9864
		300		89.46	18.64	390.60	0.9825
		305		82.85	17.70	385.17	0.9797
		310		76.51	16.77	379.59	0.9777
		315		70.40	15.85	373.87	0.9766
		320		64.48	14.91	367.98	0.9762
	Traditional mode	/	284.7	110.03	21.29	406.90	/
			290	102.60	20.30	402.74	
			295	95.72	19.37	398.51	
			300	89.03	18.43	393.99	
			305	82.53	17.50	389.18	
			310	76.21	16.56	384.07	
			315	70.08	15.62	378.65	
			320	64.13	14.67	372.93	
R1234yf	Vapor–liquid two-phase expansion mode	290	/	17.67	10.43	151.67	0.9766
		295		15.18	9.31	147.86	0.9807
		300		12.71	8.12	143.87	0.9845
		305		10.42	6.94	139.65	0.9886
		310		8.18	5.70	135.21	0.9921
		315		6.06	4.44	130.49	0.9953
		320		4.07	3.14	125.47	0.9980
	Traditional mode	/	290	17.62	10.27	154.00	/
			295	15.02	9.11	149.83	
			300	12.52	7.93	145.45	
			305	10.15	6.72	140.85	
			310	7.87	5.46	136.04	
			315	5.72	4.18	130.99	
			320	3.70	2.85	125.70	
R1234ze(E)	Vapor–liquid two-phase expansion mode	290	/	25.70	12.94	172.89	0.9792
		295		22.82	11.88	169.28	0.9817
		300		20.04	10.80	165.51	0.9845
		305		17.37	9.71	161.59	0.9873
		310		14.79	8.58	157.49	0.9902
		315		12.34	7.46	153.18	0.9930
		320		9.98	6.29	148.66	0.9958
	Traditional mode	/	290	24.53	12.29	175.13	/
			295	21.57	11.18	171.29	
			300	18.71	10.06	167.27	
			305	15.98	8.93	163.06	
			310	13.35	7.76	158.66	
			315	10.86	6.59	154.05	
			320	8.45	5.36	149.24	

Based on the data listed in Table 3, it can be seen that each of these three working fluids exhibits a better thermodynamic performance when adopting the vapor–liquid two-phase expansion mode. However, the difference in thermodynamic performance between the two modes is not very large. Of the three working fluids, cis-butene has the best performance in terms of net power output and thermal efficiency, while R1234yf and R1234ze(E) have a better performance with respect to the heat exchange load of the condenser. The indexes of all three working fluids decrease with increasing condensation temperature under both working modes, the vapor–liquid two-phase expansion mode and the traditional mode.

If we compare the two working modes with respect to the same indicator of the same working fluid, it can be seen that cis-butene behaves differently from the other two working fluids. For cis-butene, the difference between the two modes with respect to net power output decreases first, and then increases with increasing condensation temperature. Meanwhile, for R1234yf and R1234ze(E), the difference increases. This trend can be seen in Figure 8a. For cis-butene, the difference in thermal efficiency remains basically unchanged at first, and then increases with the increase of condensation temperature. Meanwhile, for R1234yf and R1234ze(E), this difference increases. This can be seen in Figure 8b. For cis-butene, the difference in the heat exchange load of the condenser increases with increasing condensation temperature. Meanwhile, for R1234yf and R1234ze(E), this difference decreases. This can be seen in Figure 8c.

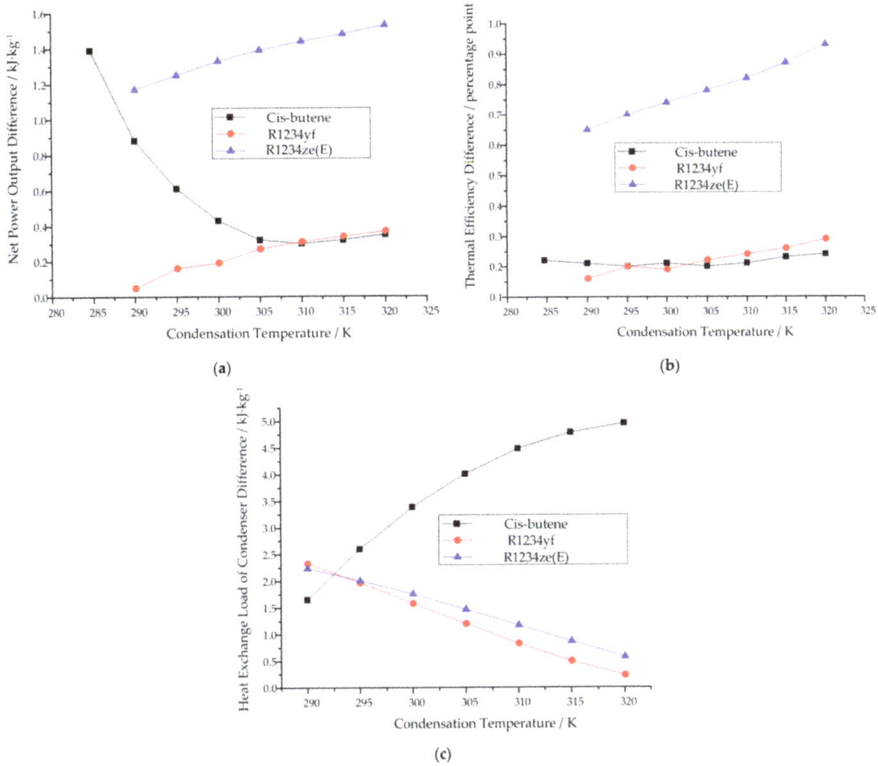

Figure 8. Variation in the difference in thermodynamic performance with increasing condensation temperature in the subcritical cycle model when the condensation temperature is known: (**a**) net power output, (**b**) thermal efficiency, (**c**) heat exchange load of the condenser.

Additionally, from Table 3, it can be seen that the vapor quality of cis-butene decreases with increasing condensation temperature. Meanwhile, for R1234yf and R1234ze(E), the vapor quality increases. The vapor quality of all three working fluids is very high. In other words, they are close to a saturated vapor state.

The reason for all of the above differences lies in the different relative positions of the condensation temperature range on the saturated vapor curves of the different working fluids. Specifically, for cis-butene, 320 K corresponds to the local minimum entropy. This factor is well illustrated and explained in Figures 6 and 7.

In summary, cis-butene can be selected as a suitable working fluid in the subcrictical cycle model when the condensation temperature is known. It is suitable for both open and closed types of heat source, with a higher cost to the heat exchanger.

3. Ideal Transcritical Cycle Model

3.1. Thermodynamic Setting and Description

Based on the calculation and analysis of the subcritical cycle model when the working conditions of the expander inlet or the condensation temperature are known, it can be seen that the differences in thermodynamic performance between the vapor–liquid two-phase expansion mode and the traditional mode are not large. In other words, the advantage represented by the tolerance of the single-screw expander for two-phase fluids has not been fully utilized under subcritical conditions.

Therefore, in this section, the transcritical cycle model is established to analyze the performance of the single-screw expander.

Similar to the settings in the previous subsection, condensation temperature is used to determine the maximum temperature in the transcritical cycle model. Figure 9 depicts the whole cycle process for cis-butene. Figure 10 depicts the cycle for R1234yf and R1234ze(E). In these two figures, both point 4 and point 5 are on the saturated vapor curve. Line 4–5 is an isentropic line. The temperature at point 4 is higher than $0.9T_c$ (critical temperature), which is the extreme temperature of the subcritical region. As mentioned before, for cis-butene, 320 K only corresponds to the local minimum point of entropy on its saturated vapor curve, as shown in Figure 9b. Meanwhile, for R1234yf and R1234ze(E), 290 K, which is their lowest condensation temperature, is higher than the temperature corresponding to their local minimum entropy.

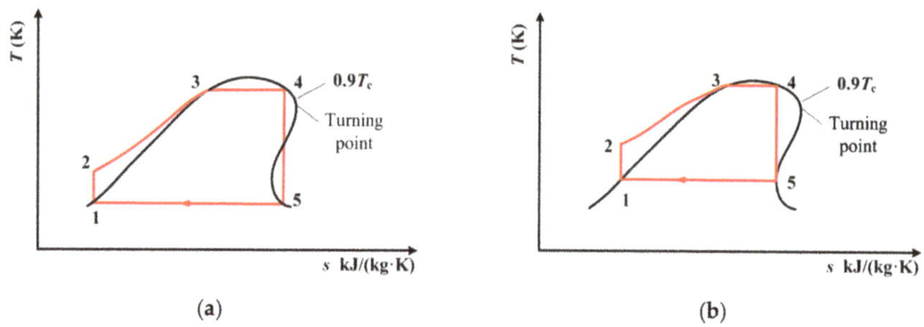

Figure 9. Transcritical ORC for cis-butene with single-screw expander when the condensation temperature is known: (**a**) 290 K, (**b**) 320 K.

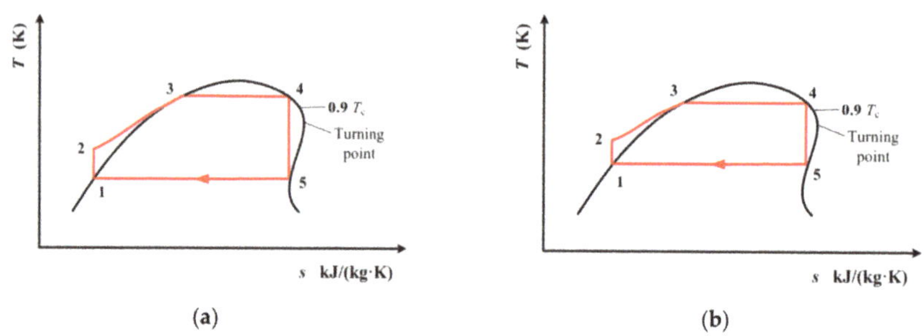

Figure 10. Transcritical ORC for R1234yf and R1234ze(E) with single-screw expander when the condensation temperature is known: (**a**) 290 K, (**b**) 320 K.

Due to the different relative positions of the condensation temperature range on the saturated vapor curves, the maximum cycle temperature that corresponds to the condensation temperature is different for the three working fluids. For cis-butene, the maximum cycle temperature that corresponds to the lowest condensation temperature is lower than the maximum cycle temperature that corresponds to the highest condensation temperature. Meanwhile, for R1234yf and R1234ze(E), their maximum cycle temperature that corresponds to the lowest condensation temperature is higher than the maximum cycle temperature that corresponds to the highest condensation temperature. Figures 9 and 10 illustrate this difference. The equations for calculating the three indicators in Figures 9 and 10 are the same as the equations used for the calculations for Figure 3.

The cycle depicted in Figure 4 is still used for the purpose of comparison. The temperature at point 4, which is the same as the temperature at the expander inlet, is still set to be the same as the temperature at the turning point. State point 4 is still on the saturated vapor curve. Point 6 is on the saturated vapor curve, and its temperature (T_6) is the same as the condensation temperature, which is varied between 290 K and 320 K.

3.2. Results and Discussion

Table 4 lists the net power output, thermal efficiency, and heat exchange load of the condenser for cis-butene, R1234yf, and R1234ze(E) in transcritical ORC when the condensation temperature is known.

Table 4. Thermodynamic performance of the three working fluids in the transcritical cycle model when the condensation temperature is known.

Working Fluid	Working Mode	Condensation Temperature T_5/K	Condensation Temperature T_6/K	Expander Inlet Temperature T_4/K	Net Power Output/kJ·kg^{-1}	Thermal Efficiency/%	Heat Exchange Load of Condenser/kJ·kg^{-1}
cis-butene (cis-2-butene)	Vapor–liquid two-phase expansion mode	284.7	/	400.00	116.37	22.26	406.46
		290		407.78	112.86	21.96	401.09
		295		411.20	107.49	21.35	395.91
		300		413.36	101.67	20.65	390.60
		305		414.72	95.69	19.90	385.17
		310		415.60	89.71	19.11	379.60
		315		416.05	83.79	18.31	373.87
		320		416.25	77.95	17.48	367.98
	Traditional mode	/	284.7	390	110.03	21.29	406.90
			290		102.60	20.30	402.74
			295		95.72	19.37	398.51
			300		89.03	18.43	393.99
			305		82.53	17.50	389.18
			310		76.21	16.56	384.07
			315		70.08	15.62	378.65
			320		64.13	14.67	372.93
R1234yf	Vapor–liquid two-phase expansion mode	290	/	350.78	23.51	13.42	151.67
		295		349.25	20.63	12.24	147.86
		300		347.59	17.74	10.98	143.87
		305		345.43	14.87	9.62	139.65
		310		343.10	11.97	8.13	135.21
		315		340.27	9.02	6.47	130.49
		320		336.94	6.00	4.56	125.47
	Traditional mode	/	290	330	17.62	10.27	154.00
			295		15.02	9.11	149.83
			300		12.52	7.93	145.45
			305		10.15	6.72	140.85
			310		7.87	5.46	136.04
			315		5.72	4.18	130.99
			320		3.70	2.85	125.70
R1234ze(E)	Vapor–liquid two-phase expansion mode	290	/	361.99	31.33	15.34	172.89
		295		360.87	28.14	14.25	169.28
		300		359.48	24.97	13.11	165.51
		305		357.91	21.84	11.91	161.59
		310		356.09	18.73	10.63	157.48
		315		353.90	15.58	9.23	153.18
		320		351.13	12.34	7.66	148.66
	Traditional mode	/	290	340	24.53	12.29	175.13
			295		21.57	11.18	171.29
			300		18.71	10.06	167.27
			305		15.98	8.93	163.06
			310		13.35	7.76	158.66
			315		10.86	6.59	154.05
			320		8.45	5.36	149.24

Based on the data listed in Table 4, it can be seen that when the condensation temperature is varied from 290 K to 320 K, the expander inlet temperature of cis-butene exceeds 130 °C (400 K). It has been mentioned that the maximum working temperature of the single-screw expander is 130 °C (400 K). Although it can work at temperatures higher than 400 K for a short time, its working life will be greatly reduced. Therefore, it is not recommended that single-screw expanders be operated at temperatures higher than 400 K. When 400 K is set as the inlet temperature of the single-screw expander, the corresponding condensation temperature will be 284.7 K. At this condensation temperature, the vapor–liquid two-phase expansion mode exhibits a better performance than the traditional mode.

The same as for the subcritical cycle model when the condensation temperature is known, all three thermodynamic indexes decrease with increasing condensation temperature under both working modes, the vapor–liquid two-phase expansion mode and the traditional mode. This trend can be seen in Figure 11. When comparing R1234yf with R1234ze(E) at the same condensation temperature, R1234ze(E) has better thermodynamic performance than R1234yf in terms of net power output and thermal efficiency. However, the heat exchange load of the condenser for R1234ze(E) is a little bit higher than that of R1234yf. The difference in thermodynamic performance for both R1234yf and R1234ze(E) between the two working modes decreases with increasing condensation temperature.

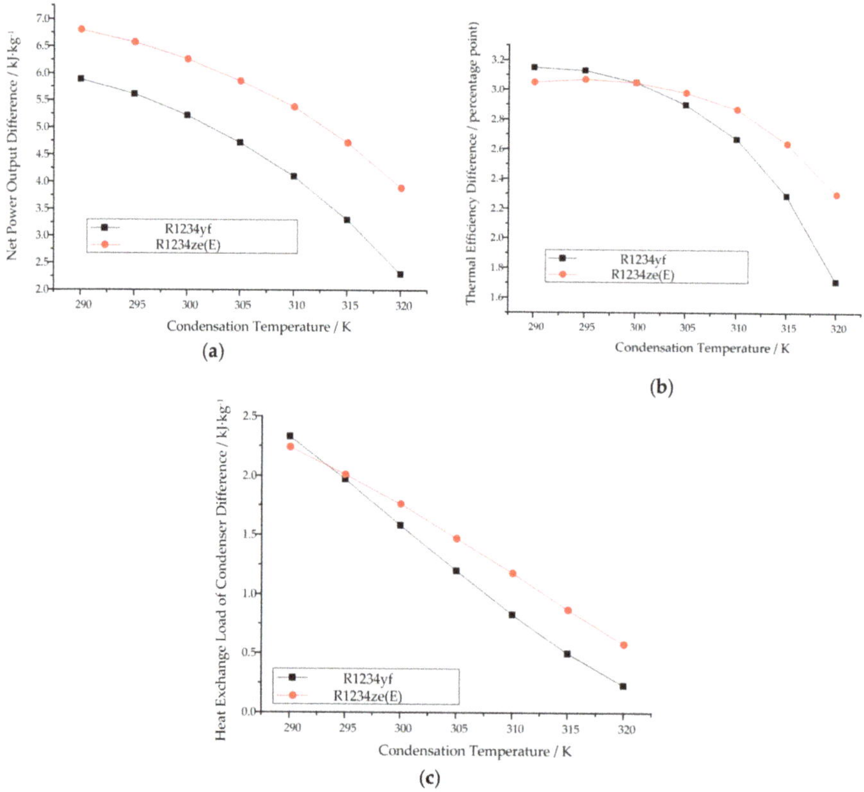

Figure 11. Variation in the difference in thermodynamic performance with increasing condensation temperature in the transcritical cycle model when the condensation temperature is known: (**a**) net power output, (**b**) thermal efficiency, (**c**) heat exchange load of the condenser.

In summary, in comparison with cis-butene, R1234yf and R1234ze(E) are more suitable for working in the transcritical cycle. Moreover, R1234ze(E) can be selected as a suitable working fluid in the transcritical cycle model when the condensation temperature is known. It is suitable for both open and closed types of heat source, with a slightly higher cost in the heat exchanger.

3.3. Comparisons of Subcritical and Transcritical Cycle Models

If we compare the thermodynamic performance of the same working fluid (R1234yf or R1234ze(E)) when working in the subcritical and transcritical cycle models at the same condensation temperature, it can be obviously found that the thermodynamic performance in the transcritical cycle model is better than that in the subcritical cycle. For both R1234yf and R1234ze(E), the difference in their thermodynamic performances between the subcritical and transcritical cycle models decreases with increasing condensation temperature. Figure 12 depicts this trend. As for cis-butene, its maximum working temperature is 400 K, which corresponds to a condensation temperature of 284.7 K. At this temperature, the thermodynamic performance of cis-butene in the transcritical cycle model is better than that in the subcritical cycle model.

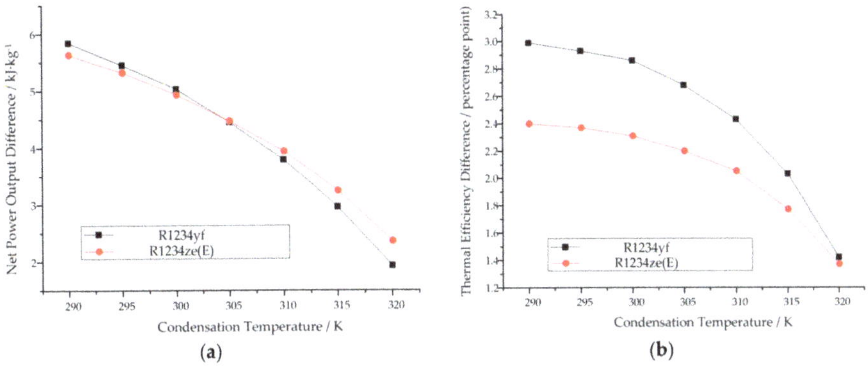

Figure 12. Thermodynamic performance difference variation between subcritical and transcritical cycle with increase of condensation temperature: (a) Net power output, (b)Thermal efficiency.

4. Analysis of the Subcritical and Transcritical Cycles Considering the Isentropic Efficiency of the Expander

4.1. Thermodynamic Setting and Description

The above two sections analyzed the ideal subcritical and ideal transcritical cycle models. They are based on the isentropic efficiency of the single-screw expander being 100%. However, in practical applications, the isentropic efficiency of different expanders needs to be taken into account. Therefore, in this section, the subcritical and transcritical cycles are analyzed considering the isentropic efficiency of the expanders. Figure 13 depicts the subcritical and transcritical cycles of cis-butene when considering the isentropic efficiency of the expander. Figure 14 depicts the subcritical and transcritical cycles of R1234yf and R1234ze(E) when considering the isentropic efficiency of the expander. In these figures, the blue dotted lines represent the expansion processes when considering the isentropic efficiency of the expander.

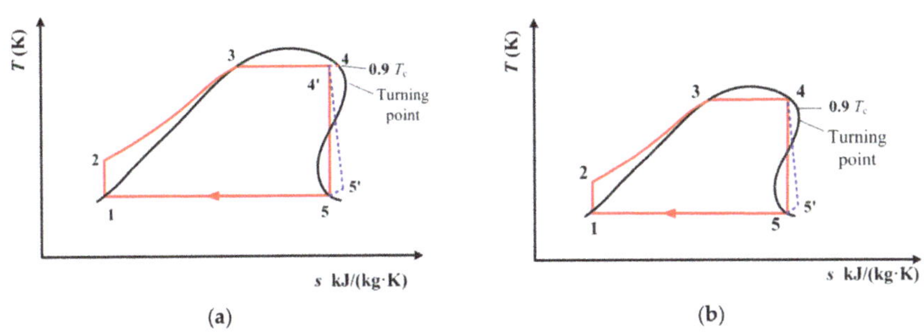

Figure 13. (a) Subcritical and (b) transcritical cycles of cis-butene when considering the isentropic efficiency of the expander.

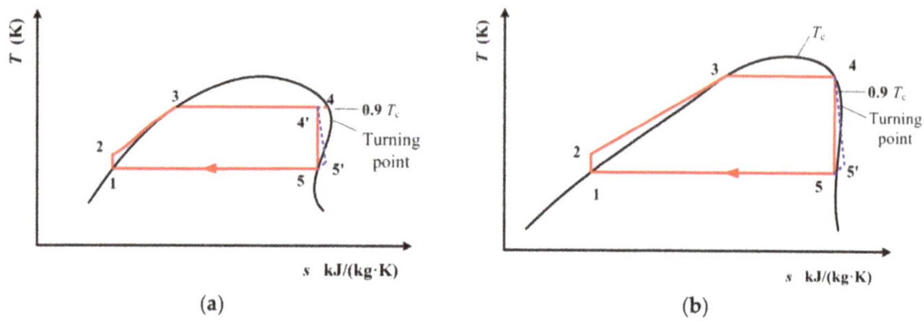

Figure 14. (**a**) Subcritical and (**b**) transcritical cycles of R1234yf and R1234ze(E) when considering the isentropic efficiency of the expander.

In Figures 13 and 14, isentropic efficiency is calculated by

$$\eta_{ex} = \frac{h_4 - h_{5'}}{h_4 - h_5}, \tag{11}$$

First of all, the isentropic efficiencies of different types of expander need to be determined. Nowadays, screw expanders show a much greater technical maturity than scroll and piston expanders [13]. The internal efficiency of single-screw expanders has exceeded 50%, and the maximum is about 65% [21,48]. Therefore, 65% is used for the analysis of single-screw expanders in this section. Table 5 lists the range of net work output values for different working fluids when taking the 65% isentropic efficiency of the single-screw expander into account. Based on the net work output range listed in Table 5, the type of expander used to perform the the comparison with single-screw expander can be determined. According to [9], piston expanders are suitable for cis-butene(cis-2-butene) and can be used for the purpose of comparison with single-screw expander. Meanwhile, scroll expanders are suitable for R1234yf and R1234ze(E). In the work conducted by Sapin et al., the efficiency of the reciprocating-piston expander ranges from 48% to 68% [49]. Therefore, 62% is adopted as the isentropic efficiency for the piston expander [50]. For scroll expanders, the maximum isentropic efficiency has reached up to 80% [51]. In this sections, we have adopted 68% as its isentropic efficiency when used in ORC [52].

Table 5. Range of net work output of different working fluids when taking the 65% isentropic efficiency of the single-screw expander into account.

Working Fluid	Working Condition	Minimum Net Work Output/kW	Maximum Net Work Output/kW
cis-butene (cis-2-butene)	Subcritical cycle	50.67	75.64
	Transcritical cycle	41.91	72.42
R1234yf	Subcritical cycle	3.90	15.28
	Transcritical cycle	2.65	11.49
R1234ze(E)	Subcritical cycle	8.02	20.36
	Transcritical cycle	6.49	16.71

4.2. Results and Discussion

Tables 6–8 show the comparisons of the thermodynamic performance of the three working fluids with and without consideration of the isentropic efficiency of the expander.

Table 6. Comparison of thermodynamic performance of cis-butene with and without consideration of the isentropic efficiency of the expander.

Working Condition	T_1	Ideal Isentropic Expansion				Considering Isentropic Efficiency of Expander			
		Net Power Output/kJ·kg^{-1}	Heat Exchange Load of Condenser/kJ·kg^{-1}	Thermal Efficiency/%	T_5/K	Net Power Output /kJ·kg^{-1}	Heat Exchange Load of Condenser/kJ·kg^{-1}	Thermal Efficiency/%	$T_{5'}$/K
Single-screw expander working in subcritical cycle	283.1	113.88	408.07	21.82	283.1	74.02	450.94	14.18	311.56
	290	103.48	401.09	20.51	290	67.26	440.30	13.33	315.46
	295	96.33	395.91	19.57	295	62.62	432.60	12.72	318.43
	300	89.46	390.60	18.64	300	58.15	424.85	12.11	321.5
	305	82.85	385.17	17.70	305	53.85	417.10	11.51	324.7
	310	76.51	379.60	16.77	310	49.73	409.24	10.90	327.96
	315	70.40	373.87	15.85	315	45.76	401.33	10.30	331.32
	320	64.48	367.98	14.91	320	41.91	393.32	9.69	334.78
Single-screw expander working in transcritical cycle	283.1	113.88	408.07	21.82	283.1	74.02	450.94	14.18	311.56
	290	112.86	401.09	21.96	290	73.36	444.61	14.27	318.2
	295	107.48	395.91	21.35	295	69.86	437.81	13.88	321.7
	300	101.67	390.60	20.65	300	66.08	430.60	13.42	325.05
	305	95.67	385.17	19.90	305	62.19	423.19	12.93	328.4
	310	89.71	379.60	19.11	310	58.31	415.55	12.42	331.74
	315	83.78	373.87	18.31	315	54.46	407.76	11.90	335.11
	320	77.94	367.98	17.48	320	50.66	399.81	11.36	338.54
piston expander woking in traditional cycle	283.1	112.40	408.10	21.59	283.12	69.69	453.70	13.39	313.34
	290	102.60	402.74	20.30	291.08	63.61	444.59	12.59	318.19
	295	95.72	398.51	19.37	296.68	59.35	437.73	12.01	321.65
	300	89.03	393.99	18.43	302.15	55.20	430.64	11.43	325.08
	305	82.53	389.18	17.50	307.49	51.17	423.33	10.85	328.48
	310	76.21	384.07	16.56	312.73	47.25	415.77	10.27	331.87
	315	70.08	378.65	15.62	317.86	43.45	407.97	9.68	335.24
	320	64.13	372.93	14.67	322.89	39.76	399.92	9.10	338.6

Table 7. Comparison of the thermodynamic performance of R1234yf with and without consideration of the isentropic efficiency of the expander.

Working Condition	T_1	Ideal Isentropic Expansion				Considering Isentropic Efficiency of Expander			
		Net Power Output /kJ·kg^{-1}	Heat Exchange Load of Condenser/kJ·kg^{-1}	Thermal Efficiency/%	T_5/K	Net Power Output/kJ·kg^{-1}	Heat Exchange Load of Condenser/kJ·kg^{-1}	Thermal Efficiency/%	$T_{5'}$/K
Single-screw expander working in subcritical cycle	290	17.68	151.67	10.44	290	11.49	158.77	6.79	297.07
	295	15.18	147.86	9.31	295	9.87	154.01	6.05	300.97
	300	12.71	143.87	8.12	300	8.26	149.11	5.28	304.94
	305	10.42	139.65	6.94	305	6.77	144.00	4.51	308.97
	310	8.18	135.21	5.70	310	5.32	138.68	3.71	313.06
	315	6.07	130.49	4.44	315	3.95	133.11	2.89	317.22
	320	4.07	125.47	3.14	320	2.65	127.25	2.04	321.44
Single-screw expander working in transcritical cycle	290	23.51	151.67	13.42	290	15.28	161.55	8.72	299.85
	295	20.63	147.86	12.24	295	13.41	156.61	7.96	303.5
	300	17.74	143.87	10.98	300	11.53	151.49	7.14	307.2
	305	14.87	139.65	9.62	305	9.67	146.09	6.26	310.9
	310	11.97	135.21	8.13	310	7.78	140.47	5.29	314.66
	315	9.02	130.49	6.47	315	5.86	134.50	4.20	318.41
	320	6.00	125.47	4.56	320	3.90	128.15	2.97	322.18
Scroll expander working in traditional cycle	290	17.62	154.00	10.27	292.31	11.98	160.52	6.98	298.82
	295	15.02	149.83	9.11	296.90	10.21	155.45	6.20	302.37
	300	12.52	145.45	7.93	301.49	8.51	150.21	5.39	305.99
	305	10.15	140.85	6.72	306.10	6.90	144.76	4.57	309.67
	310	7.87	136.04	5.47	310.73	5.35	139.14	3.72	313.47
	315	5.72	130.99	4.18	315.42	3.89	133.29	2.85	317.38
	320	3.70	125.70	2.86	320.19	2.52	127.21	1.94	321.41

Table 8. Comparison of thermodynamic performance of R1234ze(E) with and without consideration of the isentropic efficiency of the expander.

Working Condition	T_1	Ideal Isentropic Expansion				Considering Isentropic Efficiency of Expander			
		Net Power Output/kJ·kg^{-1}	Heat Exchange Load of Condenser/kJ·kg^{-1}	Thermal Efficiency/%	T_5/K	Net Power Output/kJ·kg^{-1}	Heat Exchange Load of Condenser/kJ·kg^{-1}	Thermal Efficiency/%	$T_{5'}$/K
Single-screw expander working in subcritical cycle	290	25.70	172.89	12.94	290	16.71	182.97	8.41	300.67
	295	22.82	169.28	11.88	295	14.83	178.30	7.72	304.37
	300	20.04	165.51	10.80	300	13.03	173.51	7.02	308.14
	305	17.37	161.59	9.71	305	11.29	168.59	6.31	311.97
	310	14.79	157.49	8.58	310	9.61	163.51	5.58	315.85
	315	12.34	153.18	7.46	315	8.02	158.25	4.85	319.79
	320	9.98	148.66	6.29	320	6.49	152.82	4.09	323.82
Single-screw expander working in transcritical cycle	290	31.33	172.89	15.34	290	20.36	185.58	9.97	303.43
	295	28.14	169.28	14.25	295	18.29	180.76	9.27	306.92
	300	24.97	165.51	13.11	300	16.23	175.79	8.52	310.46
	305	21.84	161.59	11.91	305	14.20	170.64	7.74	314.01
	310	18.73	157.48	10.63	310	12.17	165.31	6.91	317.6
	315	15.58	153.18	9.23	315	10.12	159.74	6.00	321.21
	320	31.33	172.89	15.34	290	20.36	185.58	9.97	303.43
Scroll expander working in traditional cycle	290	24.53	175.13	12.29	292.37	16.68	183.93	8.35	301.69
	295	21.57	171.29	11.18	297.09	14.67	179.09	7.61	305.19
	300	18.71	167.27	10.06	301.79	12.72	174.11	6.84	308.75
	305	15.98	163.06	8.93	306.46	10.87	168.95	6.07	312.32
	310	13.35	158.66	7.76	311.14	9.08	163.64	5.28	315.98
	315	10.86	154.05	6.59	315.82	7.38	158.14	4.48	319.69
	320	8.45	149.24	5.36	320.53	5.75	152.47	3.64	323.49

From Table 6, it can be seen that when considering the isentropic efficiency of the expander at condensation temperatures of 290 K and 295 K, the heat exchange load of the condenser in the traditional cycle, in which the piston expander is used, is smaller than that in the transcritical cycle in which the single-screw expander is used, but bigger than that in subcritical cycle in which the single-screw expander is used. However, at the same condensation temperature, the other aspects of thermodynamic performance of the single-screw expander working in subcritical or transcritical cycles were better than those of the piston expander working in the traditional cycle. Beyond that, at all other condensation temperatures, all aspects of the thermodynamic performance of the single-screw expander working in subcritical or transcritical cycles were better than those of the piston expander working in the traditional cycle.

From Table 7, it can be seen that when considering the isentropic efficiency of the expander, the net power output and the thermal efficiency of the single-screw expander working in the subcritical cycle are lower than those of the scroll expander working in the traditional cycle at condensation temperatures of 290 K, 295 K, 300 K, 305 K, and 310 K. At the same time, the heat exchange load of the condenser of the single-screw expander is also lower than that of the scroll expander. However, at condensation temperatures of 315 K and 320 K, the net power output and thermal efficiency of the single-screw expander working in the subcritical cycle are higher than those of the scroll expander working in the traditional cycle. Among the three working conditions, the single-screw expander working in the transcritical cycle has the highest net power output and thermal efficiency.

From Table 8, it can be seen that when considering the isentropic efficiency of the expander, the net power output and thermal efficiency of the single-screw expander working in both the subcritical and transcritical cycles are higher than those of the scroll expander working in the traditional cycle. The thermodynamic performance of the single-screw expander working in the transcritical cycle is better than that of the single-screw expander working in the subcritical cycle. At the same time, the heat exchange load of the condenser is also a little bit higher.

From above three tables, it can be seen that if the isentropic efficiency of the expander is considered, then among the subcritical cycle with the single-screw expander, the transcritical cycle with the single-screw expander, and the traditional cycle with the piston/scroll expander, the transcritical cycle has the highest net power output and thermal efficiency. At the same time, it has a slightly higher heat exchange load of the condenser. When comparing the subcritical cycle and the traditional cycle, as for cis-butene, all aspects of the thermodynamic performance of the single-screw expander are better at all condensation temperatures. For R1234yf, all aspects of the thermodynamic performance of the single-screw expander are better at a condensation temperature of 315 K. For R1234ze(E), all aspects of the thermodynamic performance of the single-screw expander are better at condensation temperatures ranging from 290 K to 310 K.

Among the three working fluids, cis-butene has the highest net power output and the highest thermal efficiency when working in the subcritical cycle. Meanwhile, R1234ze(E) may be the best choice when working in the transcritical cycle.

5. Conclusions

The organic Rankine cycle (ORC) is a popular technology used in waste heat recovery and medium- to low-temperature heat utilization. The working fluid plays a very important role in the thermodynamic cycle. The expander is a key device in ORC. In order to give full play to vapor–liquid two-phase expansion tolerance, which is an important characteristic of single-screw expanders, two ORC models—an ideal subcritical cycle model and an ideal transcritical cycle model—were established. On this basis, the isentropic efficiency of the expander was also considered for the purposes of analyzing thermodynamic performance.

Three indicators—namely, net work output, which was used to evaluate open heat source, thermal efficiency, which was used for closed heat source, and the heat exchange load of the condenser, which was used to evaluate economic performance—were chosen to analyze the performance of the ORC systems.

On the basis of our calculations and analyses, it can be seen that, without considering the isentropic efficiency of the expander, ORC systems that use single-screw expanders and undergo a vapor–liquid two-phase expansion are able to obtain higher thermal efficiency, greater net work output, and smaller heat exchange loads of the condenser. Cis-butene may be the best candidate when working in the subcritical cycle. HFO working fluids are more suitable for working in the transcritical cycle, and HFO-1234ze(E) may be the best.

When considering the isentropic efficiency of the expander, cis-butene is still the best choice for working in the subcritical cycle. Meanwhile, for the transcritical cycle, HFO-1234ze(E) may still be the best.

Author Contributions: Conceptualization, X.Z.; Data curation, Y.Z., X.Z., and M.C.; Formal analysis, X.Z. and Y.Z.; Funding acquisition, X.Z.; Methodology, X.Z.; Resources, Y.W., C.M., and J.W.; Writing—Original Draft, X.Z.; Writing—Review & Editing, J.W.

Funding: This research was funded by the National Natural Science Foundation of China (Grant No.51506001) and the Beijing Municipal Education Commission (KM201710005029). The authors gratefully acknowledge them for financial support of this work.

Conflicts of Interest: The authors declare no conflict of interest.

References

1. Imre, A.; Kustán, R.; Groniewsky, A. Thermodynamic selection of the optimal working fluid for organic Rankine cycles. *Energies* **2019**, *12*, 2028. [CrossRef]
2. Zhang, X.; He, M.; Wang, J. A new method used to evaluate organic working fluids. *Energy* **2014**, *67*, 363–369. [CrossRef]
3. Liu, B.; Chien, K.; Wang, C. Effect of working fluids on organic Rankine cycle for waste heat recovery. *Energy* **2004**, *29*, 1207–1217. [CrossRef]
4. Longo, G.; Mancin, S.; Righetti, G.; Zilio, C. Saturated vapour condensation of R134a inside a 4 mm ID horizontal smooth tube: Comparison with the low GWP substitutes R152a, R1234yf and R1234ze(E). *Int. J. Heat Mass Transf.* **2019**, *133*, 461–473. [CrossRef]
5. Ghafri, S.; Rowland, D.; Akhfash, M.; Arami-Niya, A.; Khamphasith, M.; Xiong, X.; Tsuji, T.; Tanaka, Y.; Seiki, Y.; May, E.; et al. Thermodynamic properties of hydrofluoroolefin (R1234yf and R1234ze(E)) refrigerant mixtures: Density, vapour-liquid equilibrium, and heat capacity data and modeling. *Int. J. Refrig.* **2019**, *98*, 249–260. [CrossRef]
6. Yang, J.; Ye, Z.; Yu, B.; Ouyang, H.; Chen, J. Simultaneous experimental comparison of low-GWP refrigerants as drop-in replacements to R245fa for Organic Rankine cycle application: R1234ze(Z), R1233zd(E), and R1336mzz(E). *Energy* **2019**, *173*, 721–731. [CrossRef]
7. Li, Z.; Liang, K.; Jiang, H. Experimental study of R1234yf as a drop-in replacement for R134a in an oil-free refrigeration system. *Appl. Therm. Eng.* **2019**, *153*, 646–654. [CrossRef]
8. Zhang, X.; Cao, M.; Yang, X.; Guo, H.; Wang, J. Economic analysis of organic Rankine cycle using R123 and R245fa as working fluids and a demonstration project report. *Appl. Sci.* **2019**, *9*, 288. [CrossRef]
9. Bao, J.; Zhao, L. A review of working fluid and expander selections for organic Rankine cycle. *Renew. Sustain. Energy Rev.* **2013**, *24*, 325–342. [CrossRef]
10. Velez, F.; Segovia, J.J.; Martín, M.C.; Antolín, G.; Chejne, F.; Quijano, A. A technical, economical and market review of organic Rankine cycles for the conversion of low-grade heat for power generation. *Renew. Sustain. Energy Rev.* **2012**, *16*, 4175–4189. [CrossRef]
11. Lei, B.; Wang, W.; Wu, Y.; Ma, C.; Wang, J.; Zhang, L.; Li, C.; Zhao, Y.; Zhi, R. Development and experimental study on a single screw expander integrated into an Organic Rankine Cycle. *Energy* **2016**, *116*, 43–52. [CrossRef]
12. Dumont, O.; Parthoens, A.; Dickes, R.; Lemort, V. Experimental investigation and optimal performance assessment of four volumetric expanders (scroll, screw, piston and roots) tested in a small-scale organic Rankine cycle system. *Energy* **2018**, *165*, 1119–1127. [CrossRef]

13. Lemort, V.; Guillaume, L.; Legros, A.; Declayea, S.; Quoilin, S. A comparison of piston, screw and scroll expanders for small scale Rankine cycle systems. In Proceedings of the 3rd International Conference on Microgeneration and Related Technologies, Naples, Italy, 15–17 April 2013.
14. Goodyear, J.W. Pressure Energy Translating and Like Devices. US Patent No.2716861, 6 September 1955.
15. Zimmern, B. Worm Rotary Compressors with Liquid Joints. US Patent No.3180565, 27 April 1965.
16. Desideri, A.; Broek, M.; Gusev, S.; Lemort, V.; Quoilin, S. Experimental campaign and modeling of a low-capacity waste heat recovery system based on a single screw expander. In Proceedings of the 22nd International Compressor Engineering Conference at Purdue, West Lafayette, IN, USA, 14–17 July 2014.
17. Ziviani, D.; Groll, E.A.; Braun, J.E.; Paepe, M. Review and update on the geometry modeling of single-screw machines with emphasis on expanders. *Int. J. Refrig.* **2018**, *92*, 10–26. [CrossRef]
18. Giuffrida, A. Improving the semi-empirical modelling of a single-screw expander for small organic Rankine cycles. *Appl. Energy* **2017**, *193*, 356–368. [CrossRef]
19. Shen, L.; Wang, W.; Wu, Y.; Cheng, L.; Lei, B.; Zhi, R.; Ma, C. Theoretical and experimental analyses of the internal leakage in single-screw expanders. *Int. J. Refrig.* **2018**, *86*, 273–281. [CrossRef]
20. Shen, L.; Wang, W.; Wu, Y.; Lei, B.; Zhi, R.; Lu, Y.; Wang, J.; Ma, C. A study of clearance height on the performance of single-screw expanders in small-scale organic Rankine cycles. *Energy* **2018**, *153*, 45–55. [CrossRef]
21. Lu, Y.; He, W.; Wu, Y.; Ji, W.; Ma, C.; Guo, H. Performance study on compressed air refrigeration system based on single screw expander. *Energy* **2013**, *55*, 762–768. [CrossRef]
22. He, W.; Wu, Y.; Peng, Y.; Zhang, Y.; Ma, C.; Ma, G. Influence of intake pressure on the performance of single screw expander working with compressed air. *Appl. Therm. Eng.* **2013**, *51*, 662–669. [CrossRef]
23. Zhang, Y.; Wu, Y.; Xia, G.; Ma, C.; Ji, W.; Liu, S.; Yang, K.; Yang, F. Development and experimental study on organic Rankine cycle system with single-screw expander for waste heat recovery from exhaust of diesel engine. *Energy* **2014**, *77*, 499–508. [CrossRef]
24. Yang, K.; Zhang, H.; Song, S.; Zhang, J.; Wu, Y.; Zhang, Y.; Wang, H.; Chang, Y.; Bei, C. Performance Analysis of the Vehicle Diesel Engine-ORC Combined System Based on a Screw Expander. *Energies* **2014**, *7*, 3400–3419. [CrossRef]
25. Wajs, J.; Mikielewicz, D.; Bajor, M.; Kneba, Z. Experimental investigation of domestic micro-CHP based on the gas boiler fitted with ORC module. *Arch. Thermodyn.* **2016**, *37*, 79–93. [CrossRef]
26. Mikielewicz, D.; Mikielewicz, J. A thermodynamic criterion for selection of working fluid for subcritical and supercritical domestic micro CHP. *Appl. Therm. Eng.* **2010**, *30*, 2357–2362. [CrossRef]
27. Mikielewicz, D.; Wajs, J.; Mikielewicz, J. Alternative cogeneration thermodynamic cycles for domestic ORC. *Chem. Process. Eng.* **2018**, *39*, 75–84.
28. Desideri, A.; Gusev, S.; Broek, M.; Lemort, V.; Quoilin, S. Experimental comparison of organic fluids for low temperature ORC (organic Rankine cycle) systems for waste heat recovery applications. *Energy* **2016**, *97*, 460–469.
29. White, M.T.; Oyewunmi, O.A.; Chatzopoulou, M.A.; Pantaleo, A.M.; Haslam, A.J.; Markides, C.N. Computer-aided working-fluid design, thermodynamic optimisation and thermoeconomic assessment of ORC systems for waste-heat recovery. *Energy* **2018**, *161*, 1181–1198. [CrossRef]
30. Zhang, X.; Zhang, C.; He, M.; Wang, J. Selection and Evaluation of Dry and Isentropic Organic Working Fluids Used in Organic Rankine Cycle Based on the Turning Point on Their Saturated Vapor Curves. *J. Therm. Sci.* **2019**. [CrossRef]
31. Györke, G.; Deiters, U.K.; Groniewsky, A.; Lassu, I.; Imre, A.R. Novel Classification of Pure Working Fluids for Organic Rankine Cycle. *Energy* **2018**, *145*, 288–300. [CrossRef]
32. Quoilin, S.; Broek, M.V.D.; Declaye, S.; Dewallef, P.; Lemort, V. Techno-economic survey of Organic Rankine Cycle (ORC) systems. *Renew. Sustain. Energy Rev.* **2013**, *22*, 168–186. [CrossRef]
33. Lemmon, E.W.; Huber, M.L.; McLinden, M.O. *NIST Standard Reference Database 23: Reference Fluid Thermodynamic and Transport Properties-REFPROP*; Version 9.1; National Institute of Standard Technology: Boulder, CO, USA, 19 February 2017.
34. Cis-2-butene. Available online: https://www.chemicalbook.com/ChemicalProductProperty_EN_CB7388192.htm (accessed on 1 July 2019).
35. Zhai, H.; Shi, L.; An, Q. Influence of working fluid properties on system performance and screen evaluation indicators for geothermal ORC (organic Rankine cycle) system. *Energy* **2014**, *74*, 2–11. [CrossRef]

36. American Society of Heating, Refrigerating and Air-Conditioning Engineers, Inc. ANSI/ASHRAE Standard 34-2007. *Designation and Safety Classification of Refrigerants*. ISSN: 1041-2336. Available online: https://www.ashrae.org/File%20Library/Technical%20Resources/Standards%20and%20Guidelines/Standards%20Addenda/34_2007_ak_FINAL.pdf (accessed on 1 July 2019).
37. Intergovernmental Panel on Climate Change. Changes in atmospheric constituents and in radiative forcing. In *Climate Change 2007—The Physical Science Basis, Contribution of Working Group I to the Fourth Assessment Report of the IPCC*; Cambridge University Press: Cambridge, UK, 2007; p. 212.
38. R116 Refrigerant (Hexafluoroethane, C2F6). Available online: http://www.china-refrigerants.com/r116.htm (accessed on 1 July 2019). (In Chinese).
39. American Society of Heating, Refrigerating and Air-Conditioning Engineers, Inc. ANSI/ASHRAE Standard 34-2013. *Designation and Safety Classification of Refrigerants*. ISSN: 1041-2336. Available online: https://www.ashrae.org/technical-resources/standards-and-guidelines/standards-addenda/ansi-ashrae-standard-34-2013-designation-and-safety-classification-of-refrigerants (accessed on 1 July 2019).
40. American Society of Heating, Refrigerating and Air-Conditioning Engineers, Inc. ANSI/ASHRAE Addenda 2015 Supplement. *Designation and safety classification of refrigerants*. ISSN: 1041-2336. Available online: https://www.ashrae.org/File%20Library/Technical%20Resources/Standards%20and%20Guidelines/Standards%20Addenda/34_2013_2015Supplement_20150210.pdf (accessed on 1 July 2019).
41. Climate-Friendly Alternatives to HFCs. Available online: https://ec.europa.eu/clima/policies/f-gas/alternatives_en (accessed on 1 July 2019).
42. Sánchez, D.; Cabello, R.; Llopis, R.; Arauzo, I.; Catalán-Gil, J.; Torrella, E. Energy performance evaluation of R1234yf, R1234ze(E), R600a, R290 and R152a as low-GWP R134a alternatives. *Int. J. Refrig.* **2017**, *74*, 269–282. [CrossRef]
43. Ayachi, F.; Ksayer, E.B.; Zoughaib, A.; Neveu, P. ORC optimization for medium grade heat recovery. *Energy* **2014**, *68*, 47–56. [CrossRef]
44. Le, V.L.; Feidt, M.; Kheiri, A.; Pelloux-Prayer, S. Performance optimization of low-temperature power generation by supercritical ORCs (organic Rankine cycles) using low GWP (global warming potential) working fluids. *Energy* **2014**, *67*, 513–526. [CrossRef]
45. Yan, J.L. Thermodynamic principles and formulas for choosing working fluids and parameters in designing power plant of low temperature heat. *J. Eng. Thermophys.* **1982**, *3*, 7. (In Chinese)
46. He, C.; Liu, C.; Zhou, M. A new selection principle of working fluids for subcritical organic Rankine cycle coupling with different heat sources. *Energy* **2014**, *68*, 283–291. [CrossRef]
47. Chen, H.; Goswami, D.Y.; Stefanakos, E.K. A review of thermodynamic cycles and working fluids for the conversion of low-grade heat. *Renew. Sustain. Energy Rev.* **2010**, *14*, 3059–3067. [CrossRef]
48. Ziviani, D.; Gusev, S.; Lecompte, S.; Groll, E.A.; Braun, J.E.; Horton, W.T.; Broek, M.; De Paepe, M. Characterizing the performance of a single-screw expander in a small-scale organic Rankine cycle for waste heat recovery. *Appl. Energy* **2016**, *181*, 155–170. [CrossRef]
49. Sapin, P.; Simpson, M.; White, A.J.; Markides, C. Lumped dynamic analysis and design of a high-performance reciprocating-piston expander. In Proceedings of the 30th International Conference on Efficiency, Cost, Optimisation, Simulation and Environmental Impact of Energy Systems, San Diego, CA, USA, 2–6 July 2017.
50. Zhang, B.; Peng, X.; He, Z.; Xing, Z.; Shu, P. Development of a double acting free piston expander for power recovery in transcritical CO_2 cycle. *Appl. Therm. Eng.* **2007**, *27*, 1629–1636. [CrossRef]
51. Song, P.; Wei, M.; Shi, L.; Danish, S.N.; Ma, C. A review of scroll expanders for organic Rankine cycle systems. *Appl. Therm. Eng.* **2015**, *75*, 54–64. [CrossRef]
52. Lemort, V.; Quoilin, S.; Cuevas, C.; Lebrun, J. Testing and modeling a scroll expander integrated into an Organic Rankine Cycle. *Appl. Therm. Eng.* **2009**, *29*, 3094–3102. [CrossRef]

energies

Article

Pure and Hydrocarbon Binary Mixtures as Possible Alternatives Working Fluids to the Usual Organic Rankine Cycles Biomass Conversion Systems

Costante M. Invernizzi *,†, Abubakr Ayub †, Gioele Di Marcoberardino † and Paolo Iora †

Department of Mechanical and Industrial Engineering, University of Brescia, 25123 Brescia, Italy; a.ayub@unibs.it (A.A.); gioele.dimarcoberardino@unibs.it (G.D.M.); paolo.iora@unibs.it (P.I.)

* Correspondence: costante.invernizzi@unibs.it; Tel.: +39-030-3715-569

† Current address: Department of Mechanical and Industrial Engineering, University of Brescia, via Branze, 38, 25123 Brescia, Italy.

Received: 20 October 2019; Accepted: 23 October 2019; Published: 30 October 2019

Abstract: This study investigates the use of pure and hydrocarbons binary mixtures as potential alternatives working fluids in a usual biomass powered organic Rankine cycle (ORC). A typical biomass combined heat and power plant installed in Cremona (Italy) is considered as the benchmark. Eight pure hydrocarbons (linear and cyclic) and four binary mixtures of linear hydrocarbons were selected. The critical points of the binary mixtures at different composition were calculated using an in-house code developed in MATLAB© (R2018b) environment. Based on the critical point of a working fluid, supercritical and subcritical cycle configurations of ORC were analysed. A detailed thermodynamic comparison with benchmark cycle was carried out in view of cycle efficiency, maximum operating pressure, size of the turbine and heat exchangers. The supercritical cycles showed 0.02 to 0.03 points lower efficiency, whereas, subcritical cycles showed comparable efficiencies than that of the benchmark cycle. The cycles operating with hydrocarbons (pure and mixtures) exhibited considerably lower volume flow ratios in turbine which indicates lower turbine size. Also, size parameter of regenerator is comparatively lower due to the lower molecular complexity of the hydrocarbons. A noticeable increase in turbine power output was observed with change in composition of the iso-octane/n-octane binary mixture at the same thermodynamic efficiency.

Keywords: Rankine cycle; ORC; biomass; fluid mixtures; hydrocarbons

1. Introduction

Biomass is largely used today for electricity and heat production. According to many analysts, as biomass can be easily stored, it could balance and regulate the different demands and the availability of the variable renewable energies in a context of high renewable energies uses. Other recognised positive characteristics of the biomass are its source diversification and the possibility of its local provision [1].

In 2013 the biomass energy consumption in the EU28 was about 105 Mtoe (the 2/3 of the total renewable energy production), the 75% of which used for the heat generation, the 13% for the electricity generation and the remaining 12% reserved to the transport fuels ([2], p. 33).

A sustainable future use of biomass for energy purposes has certainly to consider, among other things, the greenhouse gas emissions (direct and indirect) and has to explore the improving of the conversion efficiencies, with an increasing combined heat and electrical power production.

Anyway, in the future years the consumption of biomass for energy production is expected to grow, in different measure, according to the different policy options that will be adopted [2].

Today, in Europe about 290–300 biomass power plants are working, many of them in CHP configurations. Typically they are based on Rankine thermodynamic engines and the heat from the combustion of solid biomass is transferred to the working fluids by heat transfer oils. The Rankine cycle, at least for electrical power sizes less than 10 MW to 20 MW, employs usually an organic fluid, mostly siloxanes today [3–5]. Siloxanes are generally considered benign fluids, with a good thermal stability [6–8].

Generally, in addition to the favourable thermodynamic properties, to a low environmental impact and to a low toxicity, the cost of the working fluid is in any case an important economic parameter for the evaluation of the investment. As, for example, according to [9], 4000 L to 7000 L of working fluid are in an engine, according to its power.

As for any organic fluids a decomposition is unavoidable when the fluid works at relatively high temperatures, with serious consequences on the performances of the engine, (see, for example [10–12]), and when "contaminants" are present (water, oxygen or, mainly, lubrication oils), periodically, in a plant, a removing of the degradation products and a refilling are required. Beyond the first charge then, the cost of the new filling fluid, despite a partial recycle, could be very high [9].

The continuous investigation of different working fluids is so generally justifiable [13–16]. In this work we considered some hydrocarbons (pure or in mixtures) as potential alternative working fluids in biomass power plants.

Among the many possible working fluids available, hydrocarbons are the less expensive ones, they are flammable but most of them are considered mildly toxic. From the environmental point of view, they are considered "natural" fluids.

From the assumption to use—without any hypothetical modifications—a typical biomass boiler available today on the market, in this paper we explored—although quite preliminarly—the thermodynamic performances of supercritical and subcritical thermodynamic cycles with some hydrocarbons and mixtures of hydrocarbons as working fluids. As a comparison, we considered the performances and the design conditions of a biomass plant in Cremona (Italy) [17].

The furnace in Cremona has a design thermal outlet power of 6034 kW and the boiler has two circuits (see Figure 1): the high temperature circuit supplies 5503 kW to a heat transfer oil at 310 °C, the low temperature circuit provides 531 kW to a second mass flow of heat transfer oil at 252 °C. The working fluid in the associated thermodynamic cycles has then a maximum operative temperature of about 250 °C to 300 °C. The thermodynamic engines operates in CHP mode with a condensation temperature of about 100 °C.

These relatively low maximum temperatures and high condensation temperature combined with the relatively low thermal power available exclude the possibility to consider as alternative working fluid carbon dioxide (another "natural" fluid) or, for example, mixture of carbon dioxide and hydrocarbons. As a matter of fact, carbon dioxide should be reserved to power plants with very high power densities and to greater maximum to minimum temperatures ratio [18–21].

Some new refrigerants now available, have acceptable thermodynamic properties, good environmental parameters and a sufficient expectable thermal stability [22,23]. But unfortunately they are still very expensive.

In the past years many authors examined and proposed different approaches to the optimal working fluid selection, for low and high source temperatures. See, for example [24–32], in which, in many cases, advanced numerical multi-objective methods are used with the aim to select fluids optimising thermodynamic efficiencies and plant costs.

In this work we considered only some hydrocarbons suitable for this specific application, with the aim (1) to analyse supercritical and subcritical (with saturated vapour) thermodynamic cycles condensing at high temperature (70 °C to 100 °C), (2) to compare the results with a "conventional" option and (3) to explore and to show the potentialities of some mixtures.

Evidently, supercritical cycles could be anyway a valid option also in different situations, but, today, only a limited number of plants use this configuration. As an example, one supercritical geothermal ORC plant (the first in Europe) was in 2012 started up in Italy [33].

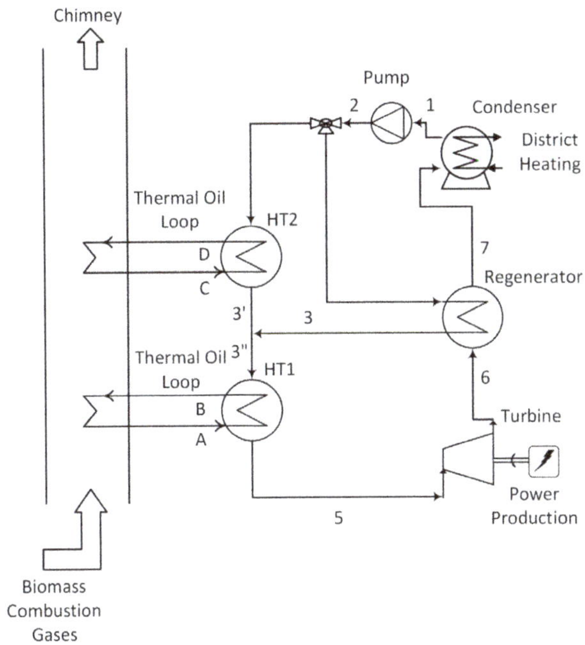

Figure 1. The reference plant scheme assumed as a reference case. Adapted from [17].

2. The Selection of The Working Fluids

In Table 1 are reported some design data for the biomass plant in Cremona (Italy). Assuming for the main cycle design parameters the values in Table 2 and MDM (octamethyltrisiloxane) as working fluid, we have simulated the thermodynamic cycle, obtaining so the reference case. All the thermodynamic calculations were carried out by Aspen Plus v9.0 using the PR EoS. For MDM cycle we obtained a gross efficiency of about 0.18: greater than the value in Table 1 (16.6%), but without considering the pressure losses. The available thermal power from the HT oil and its temperatures are fixed, according to the data in Table 1.

Table 1. Design data for the ORC plant in Cremona. The data are from [17].

Parameter	Reference Value
Thermal oil power-HT_1 loop (kW)	5503
Thermal oil temperature-HT_1 loop (inlet, T_A/outlet, T_B) (°C)	310/252
Thermal oil power-HT_2 loop (split loop) (kW)	531
Thermal oil temperature-HT_2 loop (inlet, T_C/outlet, T_D) (°C)	252/130
Overall thermal power to ORC (kW)	6034
Cooling water temperature (inlet/outlet) (°C)	65/105
Generator power (kW)	999
Gross efficiency (%)	16.6

In Figure 2 there are the critical temperatures of the two hydrocarbon families here considered as a function of the parameter of molecular complexity σ.

The parameter σ, connected with the slope of the dew line in the $T-S$ diagram, depends primarily of the molar heat capacity C_P of the working fluids and increases basically with it ([34], p. 109). An increasing of σ produces generally an isentropic expansion in the superheated vapour and, fixed a pressure expansion ratio, a reduction of the cooling of the expanding vapour (As a matter of fact, for example, for a perfect gas, an increasing of σ is strictly correlated with a reduction of the $\gamma = C_P/C_V$ and, fixed a $\partial P/P$, greater is σ, less is the cooling during an isentropic expansion $(\partial T/T = (\partial P/P)\,(1 - 1/\gamma)))$.

Table 2. Design parameters assumed for the calculations.

Parameter	Assumed Value
pressure losses	none
minimum internal temperature approach in the recuperator, $MITA_R$ (°C)	30
turbine efficiency [(a)] η_T	0.8
pump efficiency [(a)] η_P	0.75
evaporation pressure for MDM (bar)	10
condensation temperature (°C)	100
cooling water temperature (inlet/outlet) (°C)	60/90
approach temperature difference in HT_1 oil heat exchanger [(b)] (°C)	50

[(a)] including the mechanical efficiency; [(b)] for the considered super-critical cycles.

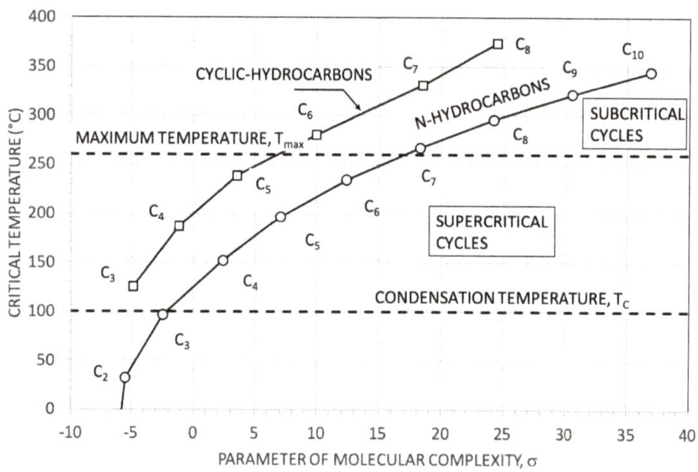

Figure 2. Critical temperature as a function of the molecular complexity parameter for linear (normal) hydrocarbons and cyclic hydrocarbons.

As, in general, for the organic molecules, an increase in the molecular complexity is accompanied by a parallel increase in the molecular weight, an increase of σ is also associated to a reduction of the isentropic enthalpy drop in turbine (with a significant reduction of the number of stage of the turbine). On the other hand, an excessive reduction of the isentropic temperature drop in turbine, requires the use of large heat recovery exchangers (the regenerator) after the turbine, before the condenser. So it is not granted working fluids with high σ are always the best option.

From this point of view, MDM is indeed, with $\sigma = 48.8$, a good example. In Figure 3a, in the $T - S$ plane, the thermodynamic cycle assumed as reference is reported. The ratio between the recovered thermal power and the input thermal power (from the heat transfer oil) $\dot{Q}_R/\dot{Q}_{in}$ is about 0.6 and the ratio $2\Delta T/\,(T_5 + T_6)$ of the corresponding isentropic expansion results equal to about 0.2 (with a pressure expansion ratio P_5/P_6 of about fifty). In Figure 3b the same thermodynamic cycles and the assumed heat sources are reported just for a clarification of the process.

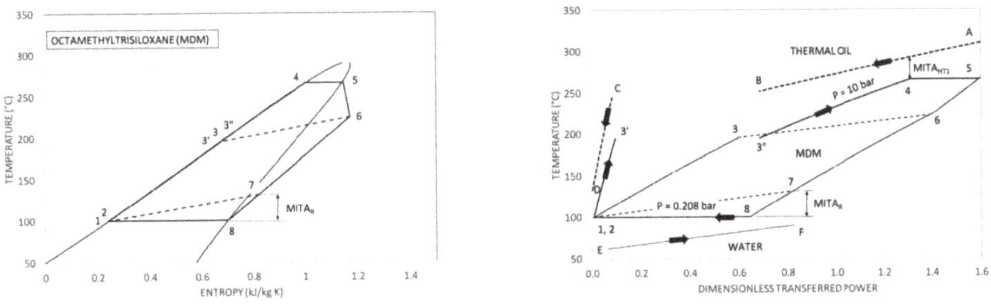

(a) (b)

Figure 3. (**a**) Reference thermodynamic cycle with MDM as working fluid in a $T-S$ plane. (**b**) The heat sources (thermal oil in HT_1 and in HT_2 loops and the cooling water) and the thermodynamic cycle in a temperature—dimensionless power plane.

In Figure 4a, for the system using MDM as working fluid in the thermodynamic cycle, the heat transfer diagram for the oil heat exchangers HT_1 and HT_2 is reported. In Figure 4b there is the corresponding diagram for the condenser.

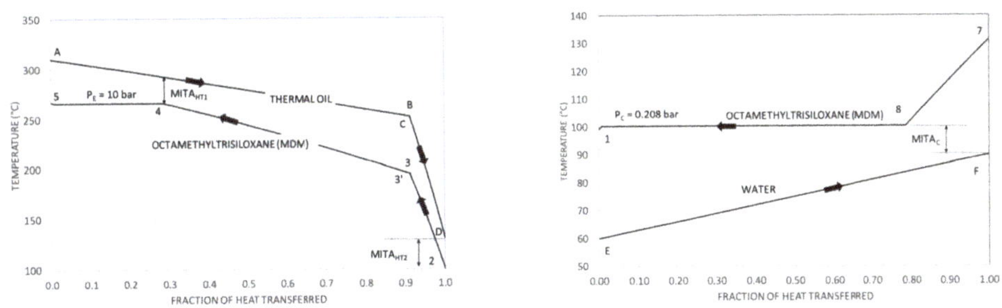

(a) (b)

Figure 4. $T-\dot{Q}$ diagrams for the reference case. (**a**) $T-\dot{Q}$ diagram for the oil heat exchangers HT_1 and HT_2. The mass flow rates of the hot oil are different in the two heat exchangers HT_1 and HT_2. So, from the point C on, the slope of the temperature changes. (**b**) $T-\dot{Q}$ diagram for the condenser.

Fixed a critical temperature, the parameter σ for cyclic hydrocarbons results appreciably lower than that of the corresponding normal (linear) hydrocarbons. For example, at $T_{cr} \approx 200$ °C, cyclo-butane has $\sigma = -1.1$ and n-pentane has σ equal to about 7.0 (see Figure 2).

On the other hand, fixed the number of carbon in the molecule, the molecular mass for cyclic and linear hydrocarbons is basically the same, and lower than that of MDM (236.53).

With the design parameters in Tables 1 and 2, the maximum, T_{max}, and the minimum T_C operating temperatures of the (pure) working fluid in the engine result equal to 260 °C and to 100 °C respectively. As the reference cycle operates in a CHP mode, in the following we assumed cooling water available between 60 and 90 Celsius degree.

For supercritical cycles, the maximum operating temperature has to be higher than the critical one; working fluids with a critical temperature greater than the assumed maximum temperature are suitable for sub-critical (with saturated vapour) thermodynamic cycles. The working fluids we preliminarily considered are then those in Table 3.

In case of mixtures in sub-critical thermodynamic cycles, $T_{max} = T_{E,dew}$ and $T_C = T_{C,dew}$: the maximum temperature equals the dew temperature at the evaporation pressure and T_C equals the dew condensation temperature.

The critical pressures P_{cr} of the cyclical hydrocarbons are appreciably greater than that of the linear hydrocarbons. For this reason, in this work, the mixtures we considered are mixture of linear hydrocarbons only (see Table 3).

The mixture n-butane/n-hexane is representative of a mixture thermodynamic suitable for supercritical cycles (given the maximum and the minimum temperatures here considered); the mixtures of iso- and n-octane, that of n-butane and n-octane and that of n-hexane and n-octane are suitable for sub-critical cycles. For these mixtures, the chosen components are those typically considered as the main constituents of commercial petrol (iso-octane, in particular, is produced and largely used to reduce the detonation of the standard petrol).

In Figure 5a,b the critical loci of the mixtures here considered are reported. In the Figure are also some $P - T$ envelops relative to some different composition.

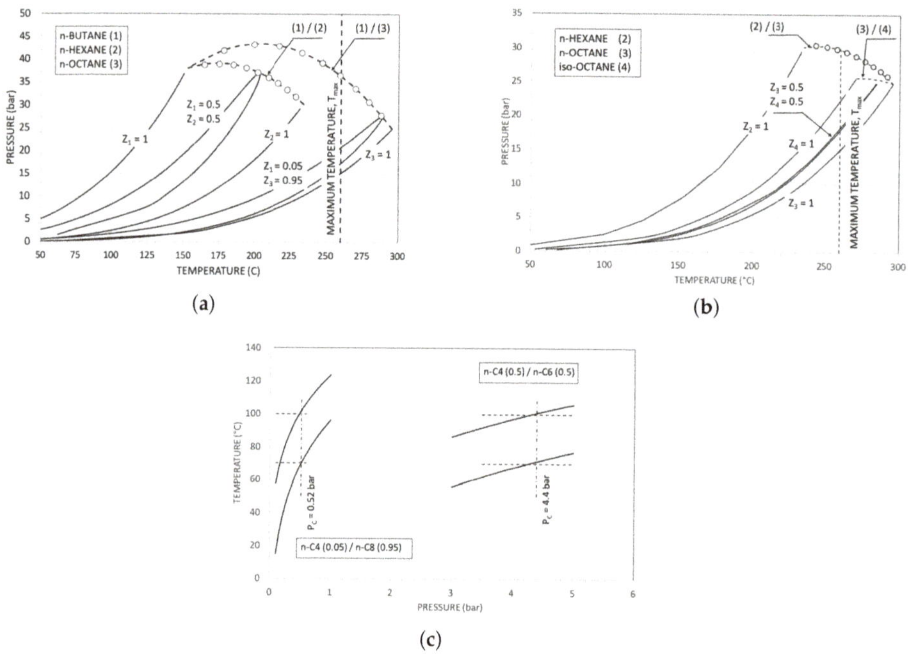

Figure 5. Critical loci for some mixtures (dotted lines). The white circle are from experimental literature data [35]. (**a**) Critical loci and some $P - T$ envelops for mixtures of n-C4/n-C6 and n-C4/n-C8. (**b**) Critical loci and $P - T$ envelops for mixtures of n-C6/n-C8 and n-C8/i-C8. (**c**) $P - T$ envelops for two mixtures around a dew temperature of 100 °C.

The critical loci are computed using an in-house code developed in MATLAB© (ver R2018b) environment with the help of INTLAB toolbox [36]. In particular, the critical point model given in [37] is used in conjunction with the PR EoS (Peng-Robinson Equation of State) [38], expressed the model in the form of non-linear system of equations with intensive variables; the roots of these equations are critical point(s) of a mixture.

The code developed in this study used the formulation of ref [38] and computes the critical points of the considered binary mixtures using interval analysis root finding technique available in INTLAB toolbox. The code requires pure fluid properties including critical points, molar composition, acentric factors and binary interaction parameter ($k_{1,2}$) corresponding to PR EoS as an input information to compute the critical points of a binary mixture. In addition, the initial interval which covers the minimum and maximum range in which the root can occur is also required.

Table 3. Some parameters for the considered fluids (pure fluids and mixtures) suitable for super-critical and sub-critical thermodynamic cycles. MDM is reported as a comparison.

Fluid (Pure Fluid or Mixture)	Critical Temperature T_{cr} (°C)	Critical Pressure P_{cr} (bar)	Vapour Pressure at 100 °C (bar)	Binary Interaction Parameter, $k_{1,2}$
n-butane, n-C4	151.97	37.96	15.33	
n-pentane, n-C5	196.55	33.7	5.91	
n-hexane, n-C6	234.45	30.25	2.45	
cyclo-propane, c-C3	125.1	55.75	36.68	
cyclo-butane, c-C4	186.85	49.8	11.27	
cyclo-pentane, c-C5	238.45	45.08	4.15	
n-C4 (0.5)/n-C6 (0.5) [(a)]	203.6	37.06	4.4 [(b)]	−0.0056
n-octane, n-C8	295.54	24.9	0.466	
iso-octane, i-C8	270.65	25.7	1.04	
i-C8 (0.5)/n-C8 (0.5) [(a)]	284.91	25.51	0.635 [(b)]	−0.00570
n-C4 (0.05)/n-C8 (0.95) [(c)]	292.02	26.2	0.505 [(b)]	0.00660
n-C6 (0.5)/n-C8 (0.5)	272.34 [(a)]	28.74 [(a)]	0.775 [(b)]	−0.01675
MDM [(d)]	291.25	14.6	0.2061	

[(a)] at an equi-molar composition ($z_1 = z_2 = 0.5$); [(b)] at the dew point; [(c)] molar composition: $z_1 = 0.05, z_2 = 0.95$; [(d)] octamethylcyclotriloxane.

In the selection of the mixtures composition we enjoined a dew temperature of around 100 °C and a "temperature glide" of about 30 °C, having us cooling water available between 60 °C and 90 °C.

Finally, a crucial parameter for the choice of a working fluid is its thermal stability. About the hydrocarbons here considered, it seems reasonable to assume a safe operative maximum temperature of 250 °C to 280 °C [39–42], according to the fixed approach temperature difference of 50 °C in the heat exchanger HT_1.

As a general rule [26,43], fixed the maximum temperature T_A of a finite thermal capacity heat source, the best thermodynamic cycle is for a working fluid with a critical temperature roughly equal to T_A. However, in our case, the source thermal power is fixed and we imposed instead the maximum cycle temperature T_{max}, so better is the thermodynamic efficiency, greater will be the useful mechanical power.

3. The Thermodynamic Results

The thermodynamic cycle efficiencies as a function of the maximum pressure P_{max} for pure linear, cyclic and for some hydrocarbon mixtures are reported in Figure 6. The efficiency value of MDM is reported as a comparison.

The super-critical cycles have generally (with the design constraints here assumed) a low thermodynamic efficiency. As a general trend, the efficiency increases with the critical temperature of the working fluid. For example, c-C3 (with a T_{cr}/T_A ratio of 0.68) has an optimal efficiency of 0.115, n-C6 (with a T_{cr}/T_A ratio of 0.87) has an optimal efficiency of 0.160.

In Figure 7a, in the $T-S$ thermodynamic plane, there are some of the considered super-critical cycles with pure working fluids. Fixed the maximum temperature T_{max}, increasing the critical temperature of the working fluid, the optimal maximum pressure approaches the critical one and the ratio W^* between the pump power $\dot{W}_P$ and the sum $\dot{W}_P + \dot{W}_T$ decreases. This is, for the super-critical cycles, the primary reason of their relatively poor efficiency. For example, set $W_0^* = 1$ for the reference MDM cycle, the cycle with c-C3 has $W^*/W_0^* \approx 7$ ($\eta = 0.115$) and the cycle with n-C6 has $W^*/W_0^* \approx 2$ ($\eta = 0.160$).

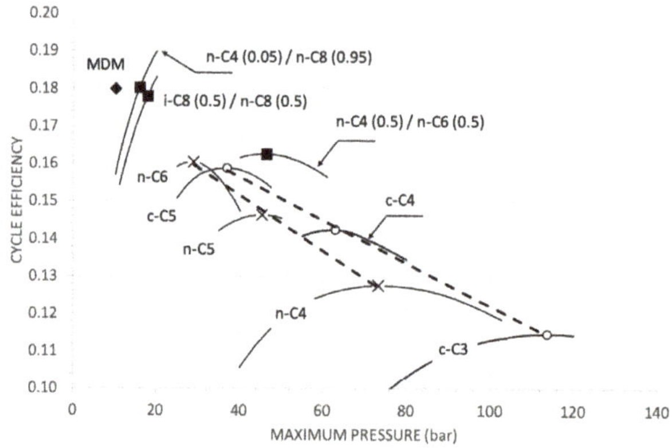

Figure 6. Thermodynamic cycle efficiency as a function of the maximum pressure.

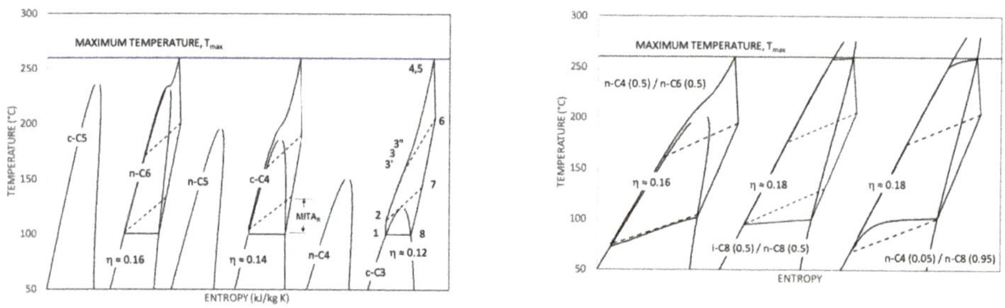

(**a**) (**b**)

Figure 7. (**a**) Configuration of some (supercritical) thermodynamic cycles in the $T - S$ plane (pure fluids); (**b**) Configuration of some thermodynamic cycles in the $T - S$ plane for hydrocarbon mixtures.

The mixture n-C4 (0.5)/n-C6 (0.5) (see Figure 5c) at a condensing pressure P_E of 4.4 bar has a dew temperature of about 100 °C and a temperature "glide" of about 30 °C. So, this particular composition results useful, in our case, to reduce the thermodynamic losses in the condenser. The corresponding thermodynamic cycles (in Figure 7b in the $T - S$ plane) has an efficiency η of about 0.164 at P_{max} of 45 bar (see Figure 6). In Figure 8a,b there are the heat transfer $T - \dot{Q}$ diagrams for the oil heat exchangers and for the condenser, respectively. The heat transfer diagram in the condenser results actually better than that of the MDM cycle, but the high temperature heat transfer diagram results worse. This fact, combined with the relatively high $W^*/W_0^* \approx 3$, gives an efficiency η anyway lower than that of the benchmarked cycle.

In Figure 6 there are also the efficiencies for two sub-critical (saturated) cycles using the mixture i-C8 /(0.5)/n-C8 (0.5) and the mixture n-C4 (0.05)/n-C8 (0.95) as working fluids. The highlighted points are those with a maximum temperature $T_{E,dew}$ of 260 °C. The corresponding cycles are reported in Figure 7b in the $T - S$ plane.

The resulting efficiencies are good (as a matter of fact comparable with that of the cycle with MDM as working fluid). The temperature profiles in the $T - \dot{Q}$ diagrams (see Figures 9 and 10) are also satisfying.

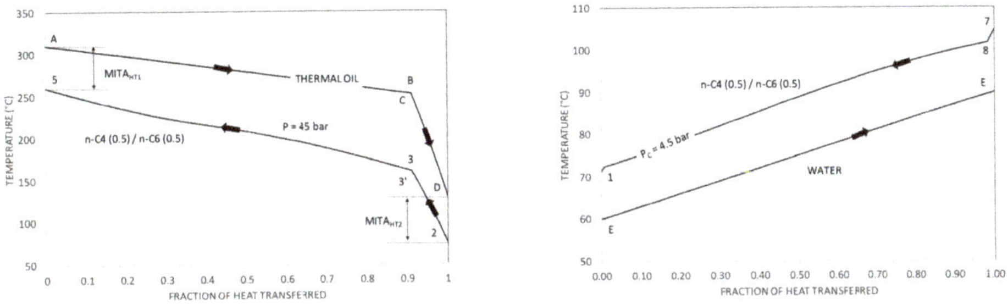

(**a**) (**b**)

Figure 8. Temperature - Thermal power exchanged ($T - \dot{Q}$ diagrams) for the the equimolar mixture n-C4/n-C6. (**a**) in the hot oil heat exchangers. (**b**) in the condenser.

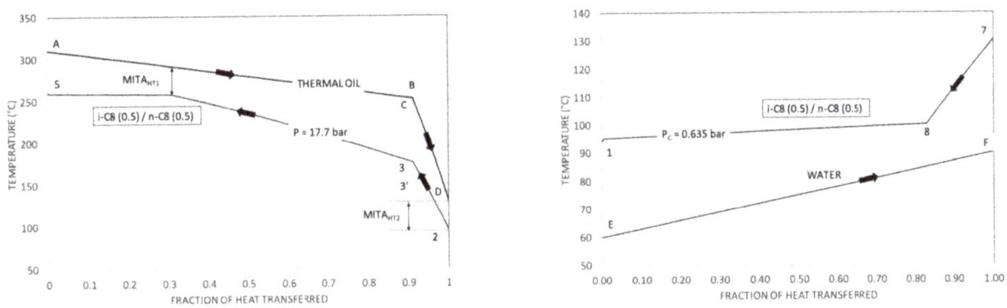

(**a**) (**b**)

Figure 9. Temperature-Thermal Power Exchanged for the cycle operating with the mixture i-C8 (0.5)/n-C8 (0.5) in the (**a**) High Temperature heat exchangers. (**b**) Condenser.

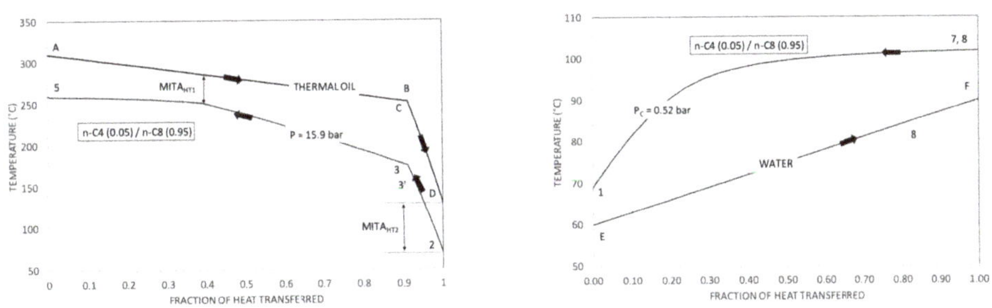

(**a**) (**b**)

Figure 10. Temperature-Thermal Power Exchanged for the cycle operating with the mixture n-C4 (0.05)/n-C8 (0.95) in the (**a**) High Temperature heat exchangers. (**b**) Condenser.

The mixture n-C6/n-C8 has a critical temperature greater than T_{max} only for a molar fraction of n-C6 lower than 0.5 and the equi-molar mixture ($T_{cr} = 272\,°C$) gives a thermodynamic cycle efficiency of 0.18 at $P_{E,dew} = 23\,\text{bar}$ and at $P_{C,dew} = 0.775\,\text{bar}$.

The mixture i-C8/n-C8 seems particularly interesting. At every composition it is like a quasi-azeotropic mixture (see Figures 5b, 9 and 11). It is a good example of the variations of thermodynamic behaviours of a mixture with two similar constituents and an example on how the main features of the thermodynamic cycles can be changed by the composition of the working fluid.

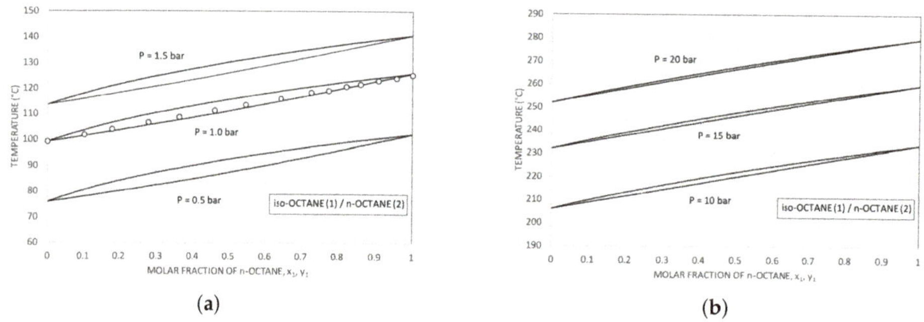

Figure 11. T – xy diagrams for the mixture i-C8 (0.5)/n-C8 (0.5) at different pressures. The white circles are from experimental literature data [44], and are reported as a comparison with the calculated values.

As explicative results, in Figure 12 the thermodynamic cycle efficiency for different mixture compositions is reported. The efficiency increases from 0.167 ($z_1 = 0$, pure iso-octane) to 0.18 ($z_1 = 1$, pure n-octane). From z_1 equal to 0 to 1, the efficiency η changes of about 7 per cent; in the interval 0.5 to 1 the variation is less than 1 per cent.

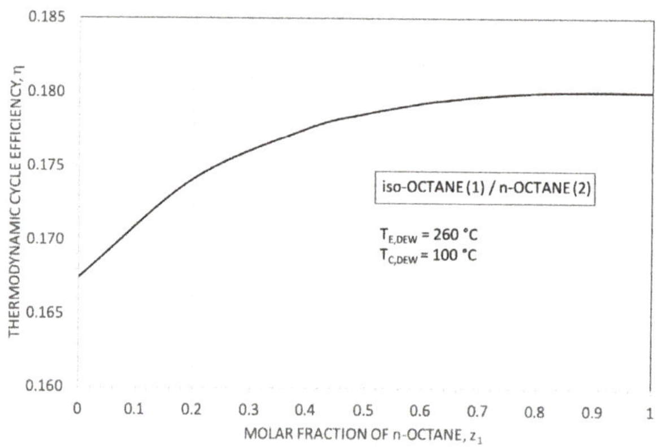

Figure 12. Thermodynamic cycle efficiency for the mixture iso-octane/n-octane as a function of the molar fraction of n-octane. The considered cycles are all subcritical with saturated vapour.

3.1. Some Results for the Heat Exchangers

For the condenser, the ratio $UA/\left(\dot{W}_T - \dot{W}_P\right)$, between UA and the net power $\dot{W}_T - \dot{W}_P$, results (for the cycles with the mixture i-C8 (0.5)/n-C8 (0.5)) equal to $0.192\,\mathrm{K}^{-1}$, as against a value of $0.170\,\mathrm{K}^{-1}$ in the case of MDM cycle (+13 per cent).

For the oil—working fluid heat exchangers (HT_1 and HT_2), the ratio $UA/\left(\dot{W}_T - \dot{W}_P\right)$ results $0.009\,16\,\mathrm{K}^{-1}$ and $0.111\,\mathrm{K}^{-1}$, respectively, as against values of $0.0114\,\mathrm{K}^{-1}$ (−20 per cent) and $0.137\,\mathrm{K}^{-1}$ (−20 per cent) for MDM.

Owing to the smaller parameter σ, the ratio $UA/\left(\dot{W}_T - \dot{W}_P\right)$ for the recuperator heat exchanger (the regenerator) results, for the thermodynamic cycle with mixture i-C8 (0.5)/n-C8 (0.5)), lower than about 40 percent compared to that of the MDM cycle ($0.0666\,\mathrm{K}^{-1}$, against $0.108\,\mathrm{K}^{-1}$).

3.2. Turbomachinery Considerations

In Figure 13a the condensation pressure (at $T_{C,dew}$ equal to 100 °C) and the evaporation temperature (at $T_{E,dew}$ equal to 260 °C) are reoprted as a function of the molar fraction of n-octane.

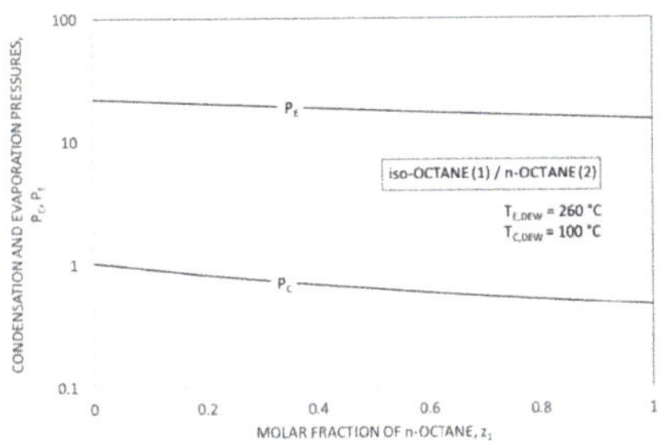

(a)

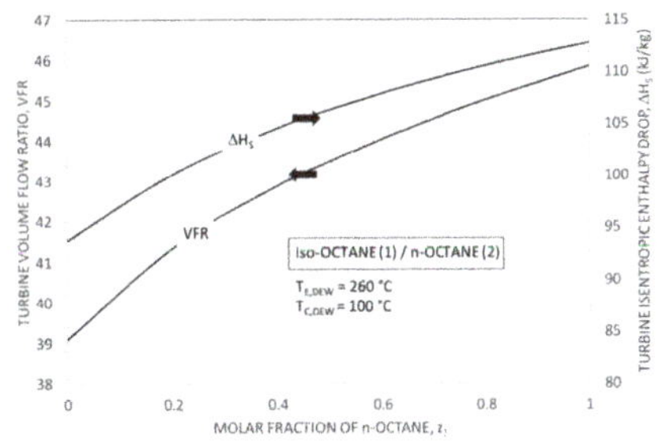

(b)

Figure 13. For a mixture of iso-octane/n-octane: (**a**) The trend of the evaporation and of the condensation pressures with the molar fraction of n-octane. (**b**) The variation of the isentropic enthalpy drop in turbine and of the isentropic volume flow ratio with the molar fraction of n-octane.

Strictly correlated with the turbine expansion pressure ratio ($P_{E,dew}/P_{C,dew}$) are the turbine volumetric flow ratio $VFR = \dot{V}_{out}/\dot{V}_{in}$ and the turbine isentropic enthalpy drop ΔH_S, see Figure 13b. The enthalpy drop and the VFR (by means of the so called size parameter SP) are closely connected with the number of turbine stages and with its rotational speed [45].

Volume flow ratios greater than about 20–30 are difficult to realise with a single turbine (axial) stage: supersonic (absolute and relative) velocities and high flaring angles increase shock and profile losses and one has to resort to a multi-stage solution. High VFR associated with low SP (less than 0.1) require multi-stages turbine at high rotating speed and, as a general trend, the result is a turbine difficult to design. For each pair of SP and VFR, exist an optimal value of the specific speed n_s and, consequently, an optimal number of revolutions N and an optimal specific diameter d_s.

For the mixture i-C8/n-C8 the VFR increases from 39 to 46, with a corresponding ΔH_S increasing from 94 kJ kg^{-1} to 112 kJ kg^{-1}. As a comparison, for the reference cycle with MDM, one obtains VFR equal to about 75 and ΔH_S equal to about 63 kJ kg^{-1}.

To summarise the results, in Table 4 are reported some significant data relating to optimal thermodynamic cycles. The considered super-critical cycles have an efficiency from 0.02 to 0.03 points lower than that of the MDM cycle. On the other hand, the turbine volume flow ratios result significantly lower than that of the MDM. Therefore an efficient single stage turbine could be designed, but, owing to the very high operating pressures, with a net power of tens of MW.

The sub-critical cycles, have thermodynamic efficiencies comparable with that of the benchmarked cycle with turbine volume flow ratios of about 30 per cent lower.

Assuming a rotating speed of 3000 rpm and an isentropic power of about 1333 kW, we estimated the approximate main characteristics of the MDM turbine. The used approximated method, assuming ideal turbines stages and real gas effects, is briefly described in [46]. Some results are in Table 5. The turbine, due to its high overall VFR, has two stages, with a rotor mean diameter of about 90 cm. The first stage is with total admission even if its h/D ratio reaches its lower limit value.

Accepting the similitude, the turbines with n-octane and iso-octane (at the same rotating speed) have roughly an isentropic power of about 5600 kW and of 7500 kW, respectively. Their main estimated geometrical data are in Table 5.

Mixtures of i-C8 and n-C8, in different composition, allow so to design turbines with significant different useful powers with practically the same thermodynamic efficiency (see Figures 12 and 13).

Table 4. Some results for supercritical and subcritical thermodynamic cycles. The results for MDM are reported as a comparison.

	Working Fluid	Maximum Pressure, P_M (bar)	Condensation Pressure, P_C (bar)	Thermodynamic Cycle Efficiency, η	Turbine Isentropic Enthalpy Drop, ΔH_S (kJ kg^{-1})	Turbine Volume Flow Ratio, $\dot{V}_{out}/\dot{V}_{in}$
Super-Critical Cycles	n-C4	70	15.6	0.128	84.77	5.12
	n-C5	40	5.99	0.145	90.59	8.28
	n-C6	30	2.49	0.160	99.83	16.94
	c-C3	110	36.8	0.115	84.56	3.01
	c-C4	60	11.35	0.142	98.42	6.04
	c-C5	40	4.16	0.158	87.13	12.262
	n-C4 (0.5)/n-C6 (0.5)	45	4.4	0.164	108.62	13.36
Sub-Critical Cycles	n-C8	15	0.47	0.180	112.2	46.00
	i-C8 (0.5)/n-C8 (0.5)	18	0.63	0.178	105.88	42.50
	n-C4 (0.05)/n-C8 (0.95)	16	0.52	0.180	112.99	43.01
	MDM	10	0.208	0.180	62.73	74.80

Table 5. Some roughly estimated design parameters for turbines at 3000 rpm. The results for MDM are assumed as a reference.

		First Stage	**Second Stage**
MDM	Number of revolutions (rpm)	3000	3000
	Volume expansion ratio	12.0	6.23
	Isentropic work (kJ kg^{-1})	31.3	31.3
	Isentropic power per stage (kW)	667	667
	Rotor tip diameter (cm)	0.455	0.487
	Rotor mean diameter (cm)	0.445	0.445
	h/D [(a)]	0.023	0.093
	SP	0.121	0.300
n-Octane	Number of revolutions (rpm)	3000	3000
	Volume expansion ratio	8.83	5.21
	Isentropic work (kJ kg^{-1})	56.1	56.1
	Isentropic power per stage (kW)	2800	2800
	Rotor tip diameter (cm)	0.611	0.648
	Rotor mean diameter (cm)	0.596	0.596
	h/D [(a)]	0.025	0.087
	SP	0.170	0.385
iso-Octane	Number of revolutions (rpm)	3000	3000
	Volume expansion ratio	9.15	4.20
	Isentropic work (kJ kg^{-1})	46.8	46.8
	Isentropic power per stage (kW)	3750	3750
	Rotor tip diameter (cm)	0.561	0.590
	Rotor mean diameter (cm)	0.545	0.545
	h/D [(a)]	0.029	0.083
	SP	0.167	0.340

[(a)] Mean rotor blade height to mean diameter ratio.

4. Conclusions

In the paper was investigated the possibility to use hydrocarbons (pure fluids or mixtures) in Rankine cycles recovering heat from biomass furnaces. As working hypothesis, we assume a standard biomass oil furnace and, as a reference thermodynamic cycle, a saturated Rankine one with MDM as working fluid. Super-critical and sub-critical thermodynamic cycles were investigated.

Notwithstanding the adopted simplifications, from the previous presented results the following conclusions can be drawn:

- the super-critical cycles, mainly owing to their high pumping power relative to the expansion useful power, have basically lower thermodynamic efficiencies than the reference cycle. In the better cases, for example, 0.16 against 0.18. The use of mixtures, allowing in principle an optimisation of the $T - \dot{Q}$ diagram in condensation, performs a bit better than the pure fluids. But, at least in the considered cases, always worse of about 1.5 points percent than the reference thermodynamic efficiency;
- the considered saturated cycles have thermodynamic efficiencies completely comparable with the reference value.

Among the many conceivable different mixtures of hydrocarbons we identifyied here mixtures of iso-octane and n-octane. These mixtures have modest temperatures "glides" at every composition and the corresponding mixture critical temperatures result always greater than the maximum assumed working temperature (260 °C). So, saturated cycles are always possible for every composition.

An interesting result is that, varying the composition the thermodynamic efficiency results in practice constant (at least starting from an n-octane molar fraction greater than 0.5). Whereas, fixed the minimum and the maximum temperatures, because of the condensation and the evaporation pressures variation, the turbine power (fixed the number of revolution) change noticeably with the composition. For example, at 3000 rpm, a two optimised stage turbine giving about 1000 kW with MDM, could produce about 4200 kW and 5600 kW with n-octane and with iso-octane respectively. The mixture composition results so an additional degree of freedom in the design of Rankine engines using organic fluids.

As for the dimension of the heat exchangers, we evaluated some results of UA per unit of produced power. Without the values of the overall heat transfer coefficients it is impossible determine the true total area A. Anyway, due to the smaller molecular complexity of the hydrocarbons compared to that of MDM, the regenerator should results with less surface.

In short, hydrocarbons as working fluids (pure or in mixtures), at least for the specific considered application, from the thermodynamic point of view, seem a good option.

Author Contributions: Conceptualization, C.M.I.; methodology, C.M.I.; software, A.A, C.M.I.; investigation, A.A., C.M.I.; writing–original draft preparation, A.A., C.M.I,; revision and final editing, G.D.M., P.I.

Funding: This research received no external funding

Conflicts of Interest: The authors declare no conflict of interest.

Abbreviations

The following abbreviations, symbols and subscripts are used in this manuscript:

A	total heat transfer area, (m^2)
CHP	Combined and Heat Power
d_s	specific diameter $\left(= D\frac{\Delta H_S^{0.25}}{\dot{V}_{out,S}^{0.5}} = \frac{D}{SP}\right)$
D	mean turbine diameter (m)
$MITA_R$	Minimum Internal Temperature Approach in the recuperator (°C)
$MITA_{HT}$	Minimum Internal Temperature Approach in the HT heat exchanger (°C)
MDM	Octamethyltrisiloxane
n_s	specific speed $\left(= \frac{2\pi N}{60}\frac{\dot{V}_{out,S}^{0.5}}{\Delta H_S^{0.75}} = \frac{2\pi N}{60}\frac{SP}{\Delta H_S^{0.5}}\right)$
N	speed revolution (rpm)
P_C	condensation pressure (bar)
P_{cr}	critical pressure (bar)
P_{max}	maximum cycle pressure (bar)
PR EoS	Peng-Robinson Equation of State
$\dot{Q}_{in}$	inlet thermal power (kW)
$\dot{Q}_R$	recovered thermal power (kW)
R	gas constant (kJ kg^{-1} K^{-1})
S	entropy (kJ kg^{-1} K^{-1})
SP	size parameter for a stage of axial turbine $\left(\dot{V}_{out,S}^{0.5}/\Delta H_S^{0.25}\right)$ (m)
T	temperature (°C or K)
T_A	maximum temperature of the heat source (thermal oil) (°C or K)
T_{cr}	critical temperature (°C or K)
T_r	reduced temperature $(= T/T_{cr})$
U	overall heat transfer coefficient (W m^{-2} K^{-1})
$\dot{V}$	volumetric flow rate (m^3 s^{-1})
VFR	isentropic flow ratio $(= \dot{V}_{out,S}/\dot{V}_{in} = \rho_{in}/\rho_{out})$
$\dot{W}$	mechanical or electrical power (kW)
$\dot{W}_S$	isentropic power (kW)
z	molar fraction
ΔH_S	isentropic turbine work (kJ kg^{-1})
ΔT	temperature difference (°C)
η_T	turbine efficiency
η_P	pump efficiency
ρ	density (kg m^{-3})
σ	parameter of molecular complexity $\left(= \frac{T_{cr}}{R}\left[\frac{dS_{SV}}{dT}\right]_{T_r=0.7}\right)$

subscripts

C	condensation
dew	dew conditions
E	evaporation
in	inlet
max	maximum
out	outlet
P	pump
S	isentropic conditions
sv	saturated vapour
T	turbine

References

1. Anonymous. State of Play o the Sustainability of Solid and Gaseous Biomass Used for Electricity, Heating and Cooling in the EU. Commission Staff Working Document. Available online: https://ec.europa.eu/energy/en/topics/renewable-energy/biomass (accessed on 8 August 2019)
2. Bogaert S.; Pelkmans L.; van den Heuvel E.; Devriendt N.; De Regel S.; Hoefnahels R.; Junginger M.; Resch G.; Liebmann L.; Mantau U.; et al. Sustainable and Optimal Use of Biomass for Energy in the EU beyond 2020—Final Report. May 2017. Available online: https://ec.europa.eu/energy/sites/ener/files/documents/biosustain_report_final.pdf (accessed on 8 August 2019).
3. Angelino, G.; Invernizzi, C. Cyclic Methylsiloxanes as Working Fluids for Space Power Cycles. *Trans. ASME J. Sol. Energy Eng.* **1993**, *115*, 130–137. [CrossRef]
4. Obernberger, I.; Biedermann, F. Combustion and Gasification of Solid Biomass for Heat and Power Production in Europe – State-of-the-Art and Relevant Future Development. In Proceedings of the 8th European Conference on Industrial Furnaces and Boilers, Vilamoura, Portugal, 25–28 March 2008; Keynote Lecture; CENERTEC: Vilamoura, Portugal, 2008; pp. 1–24.
5. Weith, T.; Heberle, F.; Preißinger, M.; Brüggermann, D. Performance of Siloxane Mixtures in a High-Temperature Organic Rankine Cycle Considering the Heat Transfer Characteristics during Evaporation. *Energies* **2014**, *7*, 5548–5565. [CrossRef]
6. Preißinger, M.; Brüggermann, D. Thermal Stability of Hexamethyldisiloxane (MM) for High-Temperature Organic Rankine Cycle (ORC). *Energies* **2016**, *9*, 183. [CrossRef]
7. Keulen, L.; Landolina, C.; Spinelli, A.; Iora, P.; Invernizzi, C.; Lietti, L.; Guardone, A. Thermal stability of hexamethyldisiloxane and octamethyltrisiloxane. *Energy Part B* **2018**, *165*, 868–876. [CrossRef]
8. Dai, X.; Shi, L.; Qian, W. Thermal stability of hexamethyldisiloxane (MM) as a working fluid for organic Rankine cycles. *Int. J. Energy Res.* **2019**, *43*, 896–904. [CrossRef]
9. Erhart, T.G.; Gölz, J.; Eicker, U.; van der Broek, M. Working Fluid Stability in Large-Scale Organic Rankine Cycle-Units Using Siloxanes—Long-Term Experiences and Fluid Recycling. *Energies* **2016**, *9*, 422. [CrossRef]
10. Invernizzi, C. M.; Bonalumi, D. Thermal stability of organic fluids for Organic Rankine Cycle systems. In *Organic Rankine Cycle (ORC) Power Systems. Technologies and Applications*; Macchi, E., Astolfi, M., Eds.; Woodhead Publishing: Duxford, UK, 2017; pp. 123–151.
11. Rajabloo, T.; Bonalumi, D.; Iora, P. Effect of a partial decomposition of the working fluid on the performances of ORC power plants. *Energy* **2017**, *133*, 1013–1026. [CrossRef]
12. Dai, X.; Shi, L.; An, Q.; Qian, W. Influence of alkane working fluid decomposition on supercritical organic Rankine cycle systems. *Energy* **2018**, *153*, 422–430. [CrossRef]
13. Angelino, G.; Colonna di Paliano, P. Multicomponent Working Fluids For Organic Rankine Cycles (ORCs). *Energy* **1998**, *23*, 449–463. [CrossRef]
14. Drescher, U.; Brüggermann, D. Fluid selection for the Organic Rankine Cycle (ORC) in biomass power and heat plants. *Appl. Therm. Eng.* **2007**, *27*, 223–228. [CrossRef]
15. Invernizzi, C.M.; Iora, P.; Bonalumi, D.; Macchi, E.; Roberto, R.; Caldera, M. Titanium tetrachloride as novel working fluid for high temperature Rankine Cycles: Thermodynamic analysis and experimental assessment of the thermal stability. *Appl. Therm. Eng.* **2016**, *107*, 21–27. [CrossRef]
16. Invernizzi, C.; Binotti, M.; Bombarda, P.; Di Marcoberardino, G.; Iora, P.; Manzolini, G. Water Mixtures as Working Fluids in Organic Rankine Cycles. *Energies* **2019**, *12*, 2629. [CrossRef]

17. Salogni, A.; Alberti, D.; Metelli, M.; Bertanzi, R. Operation and maintenance of a biomass fired—Organic Rankine Cycle—CHP plant: the experience of Cremona. *Energy Procedia* **2017**, *129*, 668–675. [CrossRef]
18. Invernizzi, C.M.; van der Stelt T. Supercritical and real gas Brayton cycles operating with mixtures of carbon dioxide and hydrocarbons. *Proc. Inst. Mech. Eng. Part A J. Power Energy* **2012**, *226*, 682–693. [CrossRef]
19. Manente, G.; Lazzaretto A. Innovative biomass to power conversion systems based on cascaded supercritical CO_2 Brayton cycles. *Biomass Bioenergy* **2014**, *69*, 155–168. [CrossRef]
20. Astolfi, M.; Lasala, S.; Macchi, E. Selection Maps For ORC And CO_2 Systems For Low-Medium Temperature Heat Sources. *Energy Procedia* **2017**, *129*, 971–978. [CrossRef]
21. Invernizzi, C.M. Prospects of Mixtures as Working Fluids in Real-Gas Brayton Cycles. *Energies* **2017**, *10*, 1649. [CrossRef]
22. Invernizzi, C. M.; Iora, P.; Preißinger M.; Manzolini, G. HFOs as substitute for R-134a as working fluids in ORC power plants: A thermodynamic assessment and thermal stability analysis. *Appl. Therm. Eng.* **2016**, *103*, 790–797. [CrossRef]
23. Huo, E.; Liu, C.; Xin, L.; Li, X.; Xu, X.; Li, Q.; Wang, S.; Dang, C. Thermal stability and decomposition mechanism of HFO-1336mzz(Z) as an environmental friendly working fluid: Experimental and theoretical study. *Int. J. Energy Res.* **2019**, 1–14. [CrossRef]
24. Papadopoulos, A.I.; Stijepovic, M.; Linke, P. On the systematic design and selection of optimal working fluids for Organic Rankine Cycles. *Appl. Therm. Eng.* **2010**, *30*, 760–769. [CrossRef]
25. Toffolo, A.; Lazzaretto, A.; Manente, G.; Paci M. A multi-criteria approach for the optimal selection of working fluid and design parameters in Organic Rankine Cycle systems. *Appl. Energy* **2014**, *121*, 219–232. [CrossRef]
26. Vivian, J.; Manente, G.; Lazzaretto A. A general framework to select working fluid and configuration of ORCs for low-to-medium temperature heat sources. *Appl. Energy* **2015**, *156*, 727–746. [CrossRef]
27. Braimakis, K.; Preißinger M.; Brüggermann, D.; Karellas, S.; Panopoulos, K. Low grade waste heat recovery with subcritical and supercritical Organic Rankine Cycle based on natural refrigerants and their binary mixtures. *Energy* **2015**, *88*, 80–92. [CrossRef]
28. Astolfi, M.; Martelli, E.; Pierobon, L. Thermodynamic and technoeconomic optimization of Organic Rankine Cycle systems. In *Organic Rankine Cycle (ORC) Power Systems. Technologies and Applications*; Macchi, E., Astolfi, M., Eds.; Woodhead Publishing: Duxford, UK, 2017; pp. 173–249.
29. Scaccabarozzi, R.; Tavano, M.; Invernizzi, C. M.; Martelli, E. Comparison of working fluids and cycle optimization for heat recovery ORCs from large internal combustion engines. *Energy* **2018**, *158*, 396–416. [CrossRef]
30. Miao, Z.; Zhang, K.; Wang M.; Xu, J. Thermodynamic selection criteria of zeotropic mixtures for subcritical organic Rankine cycle. *Energy* **2019**, *167*, 484–497. [CrossRef]
31. Yağli, H.; Koç, Y.; Koç, A.; Görgülü, A.; Tandiroğlu, A. Parametric optimization and exergetic analysis comparison of subcritical and supercritical organic Rankine cycke (ORC) for biogas fuelled combined heat and power (CHP) engine exhaust gas waste heat. *Energy* **2016**, *111*, 923–932. [CrossRef]
32. Imran, M.; Haglind, F.; Lemort, V.; Meroni, A. Optimization of organic Rankine cycle power systems for waste heat recovery on heavy-duty vehicles considering the performance, cost, mass and volume of the system. *Energy* **2019**, *180*, 229–241. [CrossRef]
33. Bonafin, J.; Di Schio, C.R.; Duvia, A. Turboden, a Presentation of Recent Worldwide Developments and Large-Scale Geothermal ORC Power-Plants. *Trans. Geotherm. Resour. Counc.* **2015**, *39*, 827–831.
34. Invernizzi, C.M. *Closed Power Cycles—Thermodynamic Fundamentals and Applications*; Springer: London, UK, 2013.
35. Hicks, C.P.; Young, C.L. The Gas-Liquid Critical Properties of Binary Mixtures. *Chem. Rev.* **1975**, *75*, 119–175. [CrossRef]
36. Rump, S.M. INTLAB—INTTerval LABoratory. In *Developments in Reliable Computing*; Csendes, T., Ed.; Springer Science+Business Media: Dordrecht, The Netherlands, 1999; pp. 77–104.
37. Heidemann, R.A.; Khalil, A.M. The calculation of critical points. *AIChE J.* **1980**, *26*, 769–779. [CrossRef]
38. Michelsen, M.L.; Heidemann, A.M. Calculation of critical points from cubic two-constant equation of state. *AIChE J.* **1981**, *27*, 521–523. [CrossRef]

39. Pasetti, M.; Invernizzi, C.M.; Iora, P. Thermal stability of working fluids for organic Rankine cycles: An improved survey method and experimental results for cyclopentane, isopentane and n-butane. *Appl. Therm. Eng.* **2014**, *73*, 764–774. [CrossRef]
40. Invernizzi, C.M.; Iora, P.; Manzolini, G.; Lasala, S. Thermal stability of n-pentane, cyclo-pentane and toluene as working fluids in organic Rankine engines. *Appl. Therm. Eng.* **2017**, *121*, 172–179. [CrossRef]
41. Dai, X.; Shi, L.; An, Q.; Qian, W. Chemical kinetics method for evaluating the thermal stability of Organic Rankine Cycle working fluids. *Appl. Therm. Eng.* **2016**, *100*, 708–713. [CrossRef]
42. Dai, X.; Shi, L.; An, Q.; Qian, W. Screening of hydrocarbons as supercritical ORCs working fluids by thermal stability. *Energy Convers. Manag.* **2016**, *126*, 632–637. [CrossRef]
43. Invernizzi, C.; Bombarda, P. Thermodynamic performance of selected HCFCS for geothermal applications. *Energy* **1997**, *22*, 887–895. [CrossRef]
44. Bromiley, E.C.; Quiggle, D. Vapor-Liquid Equilibria of Hydrocarbon Mixtures. *Ind. Eng. Chem.* **1933**, *25*, 1136–1138. [CrossRef]
45. Macchi, E.; Astolfi, M. Axial flow turbines for Organic Rankine Cycle applications. In *Organic Rankine Cycle (ORC) Power Systems. Technologies and Applications*; Macchi, E., Astolfi, M., Eds.; Woodhead Publishing: Duxford, UK, 2017; pp. 299–319.
46. Bombarda, P.; Invernizzi, C. Binary liquid metal-organic Rankine cycle for small power distributed high efficiency systems. *Proc. IMechE Part A J. Power Energy* **2015**, *229*, 192–209. [CrossRef]

energies

MDPI

Article

The Method of the Working Fluid Selection for Organic Rankine Cycle (ORC) Systems Employing Volumetric Expanders

Piotr Kolasiński

Department of Thermal Sciences, Wrocław University of Science and Technology, Wybrzeże Wyspiańskiego 27, 50-370 Wrocław, Poland; piotr.kolasinski@pwr.edu.pl; Tel.: +48-71-320-23-39

Received: 28 December 2019; Accepted: 21 January 2020; Published: 24 January 2020

Abstract: The working fluid selection is one of the most important issues faced when designing Organic Rankine Cycle (ORC) systems. The choice of working fluid is dictated by different criteria. The most important of them are safety of use, impact on the environment, and physical and chemical parameters. The type of ORC system in which the working fluid is to be used and the type of expander applied in this system is also affecting the working fluid selection. Nowadays, volumetric expanders are increasingly used in ORC systems. In the case of volumetric expanders, in addition to the aforementioned working fluid selection criteria, additional parameters are considered during the selecting of the working fluid, such as the range of operating pressures and geometric dimensions (determining the volume of working chambers) affecting the achieved power and efficiency of the expander. This article presents a method of selecting a working medium for ORC systems using volumetric expanders. This method is based on the dimensionless rating parameters applied for the comparative analysis of different working fluids. Dimensionless parameters were defined for selected thermal properties of the working fluids, namely thermal capacity, mean temperature of evaporation, mean temperature of condensation, pressure and volumetric expansion ratio, volumetric expandability, as well as the heat of preheating, vaporization, superheating, cooling, and liquefaction. Moreover, isentropic expansion work was considered as the rating parameter. In this article, in addition to the working fluid selection method, computational examples related to the selection of the working fluid for the ORC system fed by a heat source featuring specified temperatures are presented. The results of calculations of rating parameters and their comparison gave an outlook on the selection of appropriate working fluids.

Keywords: ORC; working fluid; selection method; volumetric expander; thermodynamic analysis

1. Introduction

The most important design and scientific problems connected with the Organic Rankine Cycle (ORC) systems design are suitable working fluid and expander selection [1]. Currently, there are many applicable working fluids available. An expander is often selected based on system power and its application. Two types of expansion machines (i.e., turbo and volumetric expanders) can be applied nowadays in ORC systems.

Turbo expanders are mainly applied in large power (1 MW and more) ORC systems harvesting the energy from the heat sources that have high thermal power and temperature (150 °C and more). Such heat sources (e.g., waste heat) are generated in large industrial power machinery, such as steam boilers (e.g., waste steam) or gas turbines (e.g., exhaust gases). In large power ORC systems, silicone oils (e.g., MM (hexamethyldisiloxane) or MDM (octamethyltrisoloxane)) are mainly adopted as working fluids [2].

Volumetric expanders are applied mainly in micro and low-power ORC systems, such as domestic and agriculture plants powered by the heat sources with small capacities, thermal power, and temperature (up to 150 °C) [3]. Low thermal parameters of the heat source also influence the working fluid selection. Only low-boiling working fluids are possible for application in this case. Such working fluids are refrigerants, their mixtures, and similar substances, for example, classically applied R123 and R245fa, as well as new-specially designed fluids, such as R1234yf, R1234ze, or SES36.

Volumetric expanders are a good option for systems in which low pressures and low working medium flows are expected [4]. From many types of the volumetric expanders piston, linear piston, screw, scroll, vane, and rotary lobe expanders are currently applied in ORC plants [5–21]. Volumetric expanders are often characterized by two important parameters, i.e., pressure expansion ratio (the parameter defined as the value of the working fluid pressure at the inlet to the expander in respect to the value of the working fluid pressure at the outlet of the expander—$\sigma_P = p_{in}/p_{out}$) and volumetric expansion ratio (the parameter defined as the value of the working fluid specific volume at the inlet to the expander in respect to the value of the working fluid specific volume at the outlet of the expander—$\sigma_{VEX} = v_1/v_2$). Piston expanders are a good option for ORC systems where high (up to 20 MPa in the case of the single stage expanders) inlet pressures of the working fluid are expected, as these machines feature the highest pressure expansion ratios. Piston expanders can be used in ORC systems powered by heat sources featuring fixed characteristics of temperature and thermal power output because it is advisable for these expanders to operate in dry vapor conditions (vapor quality equal to 1) in order to avoid the presence of liquid droplets in the cylinder. Screw expanders can be applied in systems utilizing the heat sources featuring floating temperature, thermal power, and output characteristic because in this type of the expanders, wet gas can be expanded without serious problems. The pressure expansion ratio of a screw expander typically ranges between 10 and 15. Scroll expanders are applied in many ORC systems. The pressure expansion ratio of a scroll expander typically ranges between 2 and 4.5. Multi-vane expanders are currently under development or during the phase of testing the prototypes. Therefore, they are applied mainly in ORC test-stands. Their design is compact, they can operate in wet gas conditions, and are relatively cheap. The pressure expansion ratio of the rotary multi-vane expanders typically ranges between five and seven. Rotary lobe expanders are also under research and development. These expanders are promising because of their advantages, such as insensibility to wet gas conditions, low operating pressures, and simple design. It is suspected that the maximum value of the pressure expansion ratio of a rotary lobe expander will reach the value of $\sigma_P = 6$. The range of working pressures, maximum (σ_{Pmax}), and typical (σ_{Pt}) pressure expansion ratios of different types of volumetric expanders are presented in Table 1.

Table 1. Range of the working pressures, maximum, and typical pressure expansion ratios of different types of volumetric machines [13–15].

Expander Type	$p_{in\,max}$ MPa	p_{out} MPa	σ_{Pmax}	σ_{Pt}
Piston (single stage)	20	0.1	200	200
Screw	1.5	0.1	15	10–15
Scroll	1.0	0.1	10	2–4.5
Rotary lobe	0.6	0.1	6	N/A
Multi-vane	0.7	0.1	10	5–7

In the case of volumetric expanders, the obtained output power is dependent on the volumetric and pressure expansion ratio. This is one of the characteristic features of volumetric machinery. This issue was treated in more detail in [3]. Each type of volumetric expander also features the optimum range of the pressure expansion ratio. It is worth noting here that research and development works on novel variable expansion ratio volumetric expanders are ongoing. The examples are scroll machines with injection ports [22–24], piston devices with controlled valves [25], and screw machines with sliding ports [26–28]. In contrast to turbo expanders, volumetric machines are able to operate under poor operating conditions (i.e., low working fluid flow rates and low pressures). The working fluid selection for ORC systems employing a volumetric expander should, therefore, be considered differently than in the case of the turbine-based ORC systems.

2. Thermodynamic Properties of Working Fluids Suitable for Application in ORC Systems Employing Volumetric Expanders

As indicated in the introduction, volumetric expanders are a good option for ORC systems supplied by low temperature (ca. 150 °C) heat sources. What is more, each type of volumetric expander feature specified the range of the pressure expansion ratio. For this reason, only selected working fluids can be applied in ORC systems employing a volumetric expander. The examples of working fluids that can be used in a low-power ORC system employing a volumetric expander are listed in Table 2 together with their basic thermal properties, which were obtained using REFPROP [28] and CoolProp [29] computer software. The working fluid class (according to the novel classification of working fluids which was proposed in [30]) and reference to the equation of state is also listed in this table for all of the working fluids. It should be mentioned here that R113, R114, R123, R124, R141b, R142b are phased out of use due to their high ODP (Ozone Depletion Potential). However, these substances are treated in the present study as the examples that can be used for further comparative analysis of other working fluids (i.e., those having similar physical properties to phased out ones).

Table 2. The working fluids suitable for small- and low-power Organic Rankine Cycle (ORC) systems with volumetric expanders [28–30].

No.	Working Fluid	Triple Point Temperature	Normal Boiling Point Temperature	Critical Point Parameters			Working Fluid Class [30]	Equation of State
		t_{trp} °C	t_{nbp} °C	t_{cr} °C	p_{cr} MPa	ρ_{cr} kg/m^3		
1	R113	−36.22	47.59	214.06	3.39	560.00	ANZCM	[31]
2	R114	−92.52	3.59	145.68	3.25	579.97	AZCM	[32]
3	R123	−107.15	27.82	183.68	3.66	550.00	ACNMZ	[33]
4	R124	−199.15	−11.96	122.28	3.62	560.00	ACNZM	[34]
5	R1234ze	−104.53	−18.95	109.37	3.63	489.24	ACNZM	[35]
6	R134a	−103.30	−26.07	101.06	4.06	512.00	ACZ	[36]
7	R152a	−118.59	−24.02	113.26	4.51	368.00	ACZ	[37]
8	R227ea	−128.60	−16.34	101.75	2.93	594.25	ANCMZ	[38]
9	R236fa	−93.63	−1.44	124.92	3.20	551.30	ACNMZ	[39]
10	R365mfc	−34.15	40.15	186.85	3.22	473.84	ANZCM	[38]
11	R245ca	−81.65	25.13	174.42	3.39	523.59	ANCMZ	[40]
12	R245fa	−102.10	15.14	154.01	3.65	516.08	ACNMZ	[41]
13	R601a	−160.50	27.83	187.2	3.38	236.00	ANCMZ	[41]
14	R141b	−103.47	32.05	204.35	4.21	458.56	ACNMZ	[41]
15	R142b	−130.43	−9.12	137.11	4.05	446.00	ACNMZ	[41]
16	R236ea	−103.15	6.19	139.29	3.5	563.00	ANZCM	[42]
17	R600a	−159.42	−11.75	134.66	3.63	225.5	ACNMZ	[43]
18	RC318	−39.80	−5.97	115.23	2.78	620.00	AZCM	[32]
19	R1234yf	−53.15	−29.45	94.7	3.38	475.55	ACNZM	[44]
20	R290	−187.63	−42.11	96.7	4.25	220.48	ACZ	[45]

Working fluids can be described by well-known thermodynamic relations and parameters (e.g., equations of state, specific heat, heat of a phase change, or the thermodynamic functions). Additionally, different parameters can be defined for their description and analysis of applicability to specified conditions. The above-mentioned thermodynamic relations are presented in the form of algebraic or differential equations, tables, graphs, software, etc. Completeness of these data in the case of the substances listed in Table 2, is very different, which does not facilitate the thermodynamic analyses. Figure 1 shows the T–s diagram for a low-boiling substance. The characteristic values of working fluid thermal properties are marked on this graph together with the ambient parameters.

The different phases of the working fluid can, therefore, be highlighted, i.e., superheated vapor, moist vapor, and a liquid. Moreover, the area of the critical point (C_r), dry saturated vapor line ($x = 1$), boiling liquid line ($x = 0$), and pinch point (ΔT_{pp}) can be marked. Working fluid pressure $p(T_a)$ at ambient temperature (T_a) typically differs from the ambient pressure (p_a). The temperature of working fluid evaporation (T_{ev}) at ambient pressure (p_a) is one of the basic quantities describing the working fluid, just like the heat of condensation $q_c(p_a)$ (see isobar-isotherm e–f in Figure 1). The critical point parameters (p_{cr}, T_{cr}) are other key parameters describing the working fluid. The value of critical pressure (p_{cr}) is a very important parameter defining the operating range of the power plant and organization of the cycle. The critical temperature (T_{cr}) determines the usefulness of a working fluid to be applied as a heat transfer fluid in a power plant. If $T_c < T_a$, it would not be possible to liquefy the vapor and pump the liquid, which are the basic processes of a vapor power plant operation. In the T–s graph, which is presented in Figure 1, the power plant cycle (see red lines in Figure 1) and heat source cooling process (see navy blue line in Figure 1) were also visualized.

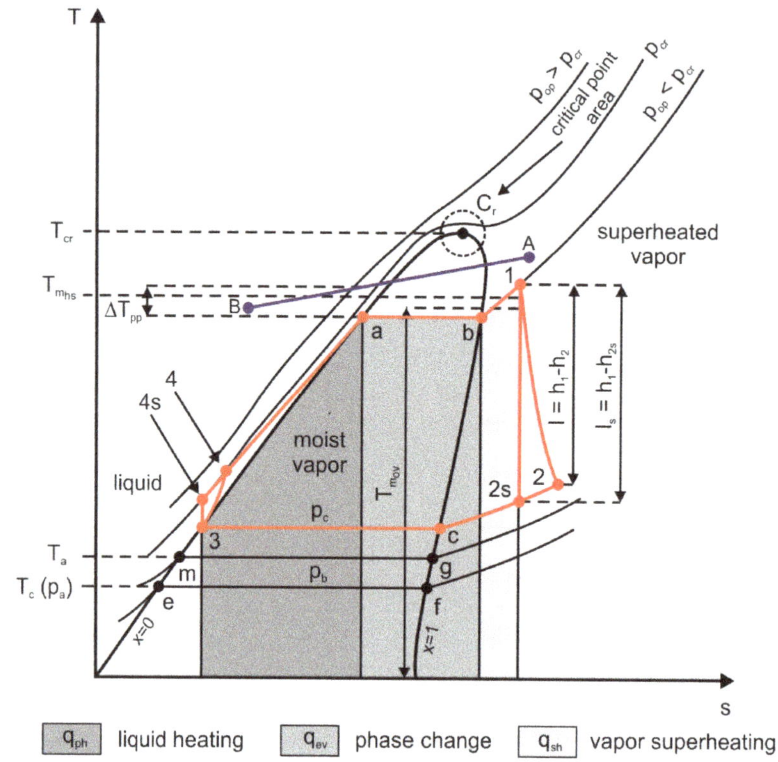

Figure 1. T–s diagram of a low-boiling working fluid.

Depending on the working fluid type, the saturation curve can have a different shape. Figure 2 shows a comparison of the saturation curves for selected low-boiling working fluids.

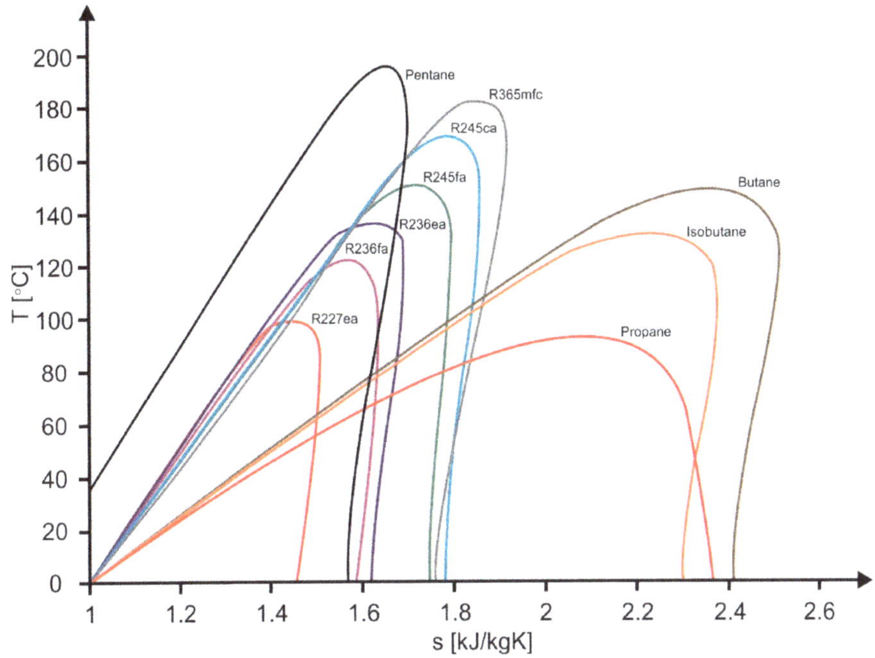

Figure 2. Saturation curves of different low-boiling working fluids.

3. The Method of Working Fluid Selection

The working fluid selection may be proceeded in many different ways depending on the accepted selection criteria. Different methods of working fluid selection are consciously developed and described in scientific papers, see [46–54]. Nowadays, environment protection is an important issue; therefore, many working fluids that were successfully applied in ORCs are being withdrawn (e.g., R11, R12, and others). The environmental impact of the working fluid should be considered during the assessment of the working fluid candidates. The thermal stability of the substance is also of great importance during the working fluid selection [54,55]. However, to fit the working fluid to the ORC, the system selection should be based mainly on the thermal properties of the working fluid [3,56,57]. What is more, the thermal properties of the working fluid should be analyzed together with the heat source characteristics. The thermal power of a heat source that is feeding the ORC system, physical state and type (e.g., gas, liquid), and physical parameters (e.g., density, viscosity, specific heat capacity) of the heat source's working fluid (e.g., waste heat carrier) have a significant influence on the process of the heat supply to the ORC system. Therefore, the shape of a curve marked as 1–2, which is visualized in Figure 3, may have a different course. For this reason, in addition to the analysis of phenomena that are occurring inside the ORC system (i.e., on low-boiling working fluid side), it is also important to analyze the phenomena that are occurring on the heat source side (i.e., related to heat source's working fluid). These problems are treated in further discussion. The starting point can be, for example, the course of the heat source cooling process. The heat source cooling process is usually isobaric; however, the course and rate of cooling can be different, and it depends on many different thermal properties of the heat source and working fluid (e.g., heat source temperature, heat capacity of the heat source, specific heat capacity of the working fluid, working fluid mass flow rate, number of heat receivers, etc.). Therefore, the different amounts of heat can be extracted from the heat source at different ranges of the temperature (this process is visualized in Figure 3). These amounts of heat can be presented graphically in a bar chart forming the heat source histogram (see Figure 4). Depending on the course and rate of the cooling process, different histograms can be obtained (see Figure 5 for histogram examples).

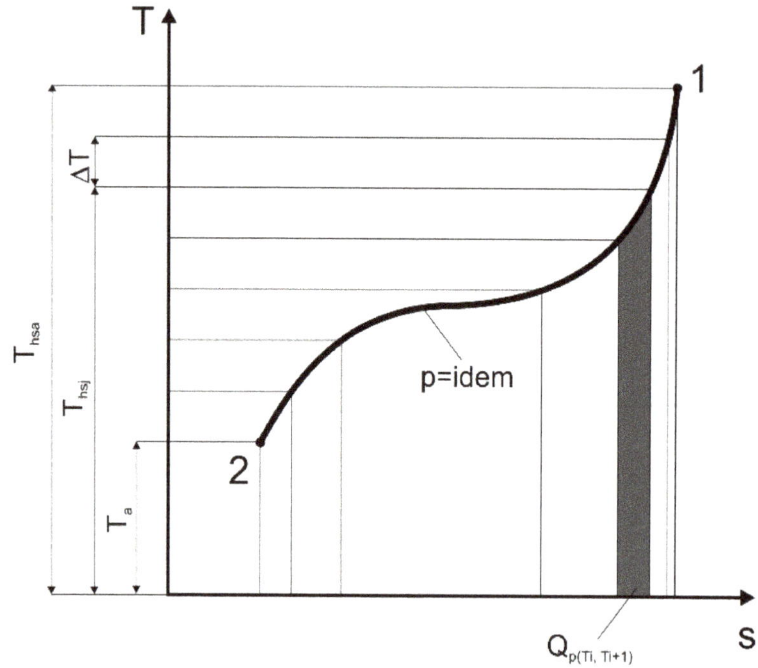

Figure 3. The process of isobaric cooling of the heat source.

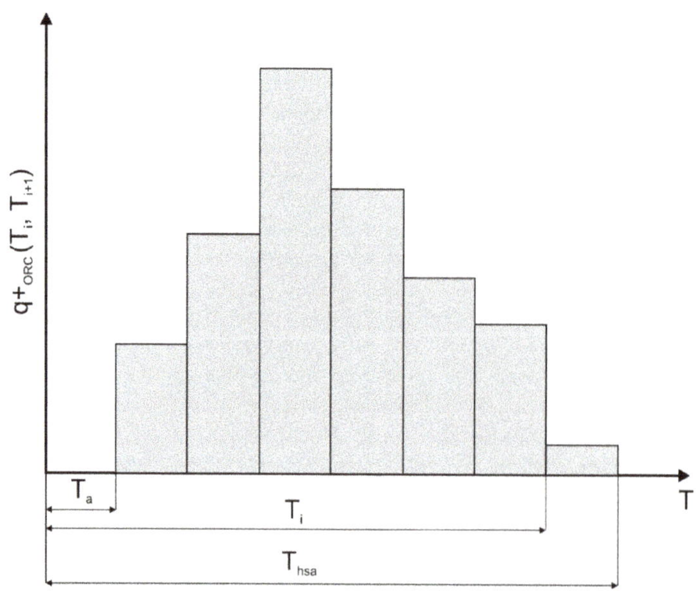

Figure 4. Example of the heat source histogram.

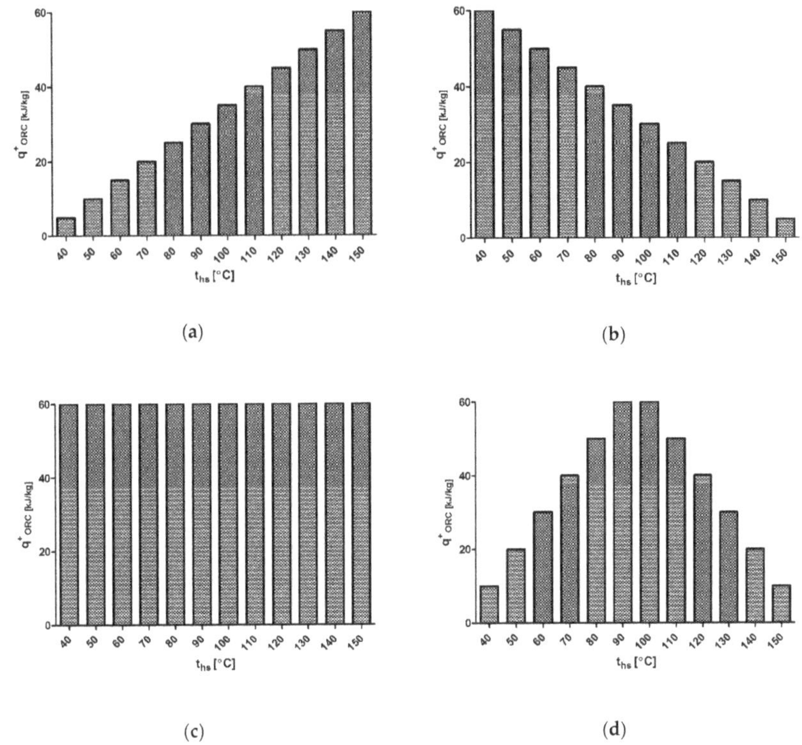

Figure 5. Histograms of different heat sources: (**a**) ascending; (**b**) descending; (**c**) constant; (**d**) parabolic.

The course of the heat source cooling process is especially important in the case of assessment of the applicability of the working fluid for an ORC system employing a volumetric expander, which is treated in more detail in the following part of this paper.

To trace the influence of the heat source cooling rate, course, and the shape of the histogram on the ORC system operating parameters and working fluid selection, the following assumptions will be accepted in further reasoning:

- The process of heat extraction from the heat source is isobaric (such as presented in Figure 3),
- Working fluid vapor condensation is isobaric and occurs at condensation pressure $p = p(T_a)$, where $T_a \approx T_c$ is ambient temperature,
- Temperature difference between vapor and coolant in the condenser is negligible,
- The pressure expansion ratio is defined as $\sigma_P = p(T_{hsa})/p(T_c)$, (where T_{hsa} is the initial temperature of the heat source),
- The expansion process in the expander is isentropic, and the value of pressure expansion ratio σ_P can be any real number,
- The amount of working fluid circulating in the cycle is equal to 1 kg,
- The power needed for pump driving is negligible,
- Expander is fed with vapor (liquid working fluid cannot be expanded).

For the above-listed assumptions, the following reasoning can be proceeded.

- If the heat source is cooled in the range of temperature $T_{hs1} \in (T_{hsa}, T_{hsa} - \Delta T)$, the obtained specific isentropic expansion work is equal to $l_{EXS1} = h_1 - h_{2s}$,

- If the heat source is cooled in the range of temperature $T_{hs2} \in (T_{hsa} - \Delta T, T_{hsa} - 2\Delta T)$, the obtained specific isentropic expansion work is equal to $l_{EXS2} = h_1' - h_{2s}'$,
- If the heat source is cooled in the range of temperature $T_{hsj} \in (T_{hsa} - (j-1)\Delta T, T_{hsa} - j\Delta T)$, the obtained specific isentropic expansion work is equal to $l_{EXSj} = h_{1j} - h_{2j}$,

where ΔT is the constant drop of the heat source temperature (it can be equal to, for example, $\Delta T = 10K$).

Therefore, the working fluid expansion can be proceeded for a given pressure expansion ratio (σ_P) value until the temperature of the working fluid is equal to (or lower) than the equilibrium temperature for the pressure $p_1 = \sigma_P \cdot p(T_a)$. Then, obtained h_j values can be visualized for a given σ_P in the $h = f(T_{hs})$ graph.

By performing the calculations for different values of the pressure expansion ratio (σ_P), a $l_{EXS} = f(\sigma_P)$ graph can be obtained. For a given heat source $l_{EXS} = f(\sigma_P)$, a graph can be plotted for different working fluids. By comparing the obtained results, the optimum pressure expansion ratio and working fluid can be selected (taking as the selection criterion the obtained values of the isentropic expansion work). What is also important, if a volumetric expander is applied in an ORC system, the obtained optimum pressure expansion ratio should fit in the range of the optimum (typical) pressure expansion ratio valid for the selected type of volumetric expander (see Table 1). Calculations can be performed for various heat source characteristics and histograms (see Figure 5).

In addition to the above-described method, the applicability of the working fluid to the ORC system can be assessed based on the comparison of the other assessment parameters. Figure 6, which is the basis for further considerations, shows the comparison of two ideal ORC power plant cycles (marked in blue and red lines) adopting two different low-boiling working fluids on the T–s plane. This graph is plotted with the assumption that both ORCs are supplied by the same heat source featuring the same heat supply temperature (T_{hs}). Moreover, it is assumed that both cycles are cooled by the same heat sink featuring the same condensation temperature (T_c), and one of the substances is treated as the reference.

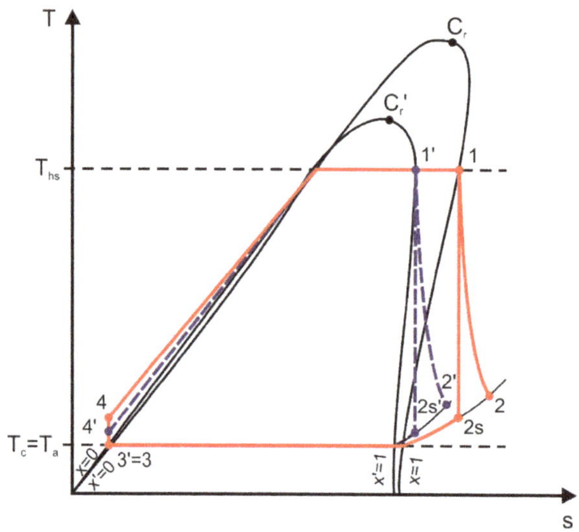

Figure 6. Comparison of two ideal ORC cycles adopting two different working fluids in the T–s plane. 4–1—isobaric heating and evaporation of the working fluid in the evaporator, 1–2s—isentropic expansion of the working fluid vapor in the expander, 1–2—polytrophic expansion of the working fluid vapor in the expander, 2–3—isobaric cooling and liquefaction of the working fluid in the condenser, 3–4—isentropic forcing of the working fluid in the pump.

Working fluids can be compared using different rating parameters or criteria. The comparison can be based on the environmental impact of the working fluids (i.e., toxicity, ODP, HGWP, etc.) or their thermal parameters. Thermal parameters of the working fluid are of great importance from the design point of view, having an impact on the ORC system power, efficiency, and quality of operation. Particularly important thermal parameters of the working fluid are:

- Thermal capacity;
- Mean temperature of evaporation;
- Mean temperature of condensation;
- Heat of evaporation (i.e., preheating, vaporization, and superheating);
- Heat of condensation (i.e., cooling and liquefying);
- Pressure expansion ratio (i.e., pressure drop of the working fluid during expansion in the expander);
- Volumetric expansion ratio (i.e., change of specific volume of the working fluid during expansion in the expander);
- Volumetric expandability (i.e., the change of specific volume of the working fluid during the evaporation).

In the following, some dimensionless rating parameters, which can be applied for the comparison of two different workings using the above-listed thermal parameters, are defined. In addition to these rating parameters, it is necessary to state the following assumptions needed to carry out the comparison (see Figure 6 for details):

- The temperature of the heat source and the heat sink is the same for both working fluids;
- Temperature of working fluid chemical decomposition is lower than the temperature of the heat source (for both of the analyzed working fluids);
- One of the working fluids is treated as the reference substance.

3.1. The Influence of the Working Fluid Thermal Capacity

The working fluid selection, which is based on the comparison of thermal capacities of the working fluids can be, for example, proceeded if a quick heat source cooling is the priority of the ORC system operation. The rating parameter can be defined by the relation

$$\psi_{TC} = \left(\frac{\dot{Q}^{+}_{ORC1}}{\dot{Q}^{+}_{ORC2}} \right)_{(T_{hs}-T_a),p} \tag{1}$$

where

$$\dot{Q}^{+}_{ORC1} = \dot{m}_{ORC1} \cdot (h_1 - h_4), \tag{2}$$

$$\dot{Q}^{+}_{ORC2} = \dot{m}_{ORC2} \cdot (h_{1'} - h_{4'}), \tag{3}$$

where $\dot{m}_{ORC1}$ is the working fluid mass flow rate in the compared cycle, $\dot{m}_{ORC2}$ is the working fluid mass flow rate in reference cycle, $\dot{Q}^{+}_{ORC1}$ is the heat extracted from the heat source by the working fluid during the evaporation in the compared cycle, $\dot{Q}^{+}_{ORC2}$ is the heat extracted from the heat source by the working fluid during the evaporation in the reference cycle.

The comparison of two different working fluids and two different ORC cycles using the ψ_{TC} parameter gives the opportunity of selecting the working fluid that features a larger thermal capacity within the fixed parameters of the heat source, i.e., the working fluid in which the application can improve the cooling rate of the heat source and thus increase the amount of heat that is extracted from the heat source.

3.2. The Influence of the Mean Temperature of the Working Fluid Evaporation

The temperature of the working fluid evaporation strongly influences the thermal efficiency of the ORC system; therefore, it is an important property from the working fluid selection point of view. The rating parameter can be defined by the equation

$$\psi_{TME} = \left(\frac{T^{mhs}_{ORC1}}{T^{mhs}_{ORC2}}\right)_{(T_{hs}-T_a),p}, \quad (4)$$

$$T^{mhs}_{ORC1} = \frac{q_{ph1}+q_{b1}+q_{sh1}}{\Delta s_1}, \quad (5)$$

$$T^{mhs}_{ORC2} = \frac{q_{ph2}+q_{b2}+q_{sh2}}{\Delta s_2}, \quad (6)$$

where T^{mhs}_{ORC1} is the mean temperature of the evaporation (and heat supply) in the compared cycle, T^{mhs}_{ORC2} is the mean temperature of the evaporation (and heat supply) in the reference cycle, q_{ph} is the heat of the working fluid preheating, q_b is the heat of the working fluid vaporization, q_{sh} is the heat of the working fluid superheating, Δs is the change of the working fluid specific entropy during the preheating, evaporation, and superheating.

The comparison of two different working fluids and two different ORC cycles using the ψ_{TME} parameter gives the opportunity of selecting the working fluid that would enable harvesting of heat from the heat source at the highest possible mean heat extraction temperature T_{mhs} (i.e., the working fluid in which the application would improve the ORC system efficiency).

The process of the working fluid evaporation is composed of the consecutive processes of preheating, boiling, and superheating. The amount of heat needed to be extracted from the heat source to preheat the liquid working fluid to the starting point of the evaporation (see point a in Figure 1), to evaporate the liquid working fluid in the range of $0 < x < 1$ (see line between points a and b in Figure 1) and to superheat the vapor of the working fluid (see line between points b and 1 in Figure 1) at a fixed heat source temperature (T_{hs}) may vary for different working fluids.

In some ORCs, preheaters (regenerators) or coolers are applied (in the case of steam power plants also super heaters); therefore, the rating parameters should also be defined for these processes.

The rating parameter for preheating of the working fluid can be defined by the equation

$$\psi_{PH} = \left(\frac{q_{PH1}}{q_{PH2}}\right)_{(T_{hs}-T_a),p} \quad (7)$$

where q_{PH1} is the amount of heat needed to be extracted from the heat source to preheat the working fluid in the compared cycle, q_{PH2} is the amount of heat needed to be extracted from the heat source to preheat the working fluid in the reference cycle.

The rating parameter for vaporization of the working fluid can be defined by the equation

$$\psi_{V} = \left(\frac{q_{V1}}{q_{V2}}\right)_{(T_{hs}-T_a),p} \quad (8)$$

where q_{V1} is the amount of heat needed to be extracted from the heat source to vaporize the working fluid in the compared cycle, q_{V2} is the amount of heat needed to be extracted from the heat source to vaporize the working fluid in the reference cycle.

The rating parameter for superheating of the working fluid can be defined by the equation

$$\psi_{SH} = \left(\frac{q_{SH1}}{q_{SH2}}\right)_{(T_{hs}-T_a),p} \quad (9)$$

where q_{SH1} is the amount of heat needed to be extracted from the heat source to superheat the working fluid in the compared cycle, q_{SH2} is the amount of heat needed to be extracted from the heat source to superheat the working fluid in the reference cycle.

The comparison of the different working fluids and different ORC cycles using ψ_{PH}, ψ_{V}, and ψ_{SH} rating parameters gives the opportunity of selecting the working fluid that would limit the amount of heat needed to be extracted from the heat source to preheat, evaporate, and eventually superheat the working fluid and thus to minimize the heat transfer area and dimensions of the applied preheater, evaporator, and superheater.

3.3. *The Influence of the Mean Temperature of Working Fluid Condensation*

The temperature of condensation also strongly influences the thermal efficiency of the ORC system; therefore, it is an important parameter from the working fluid selection point of view. The rating parameter can be defined by the equation

$$\psi_{TMC} = \left(\frac{T^{mhc}_{ORC1}}{T^{mhc}_{ORC2}}\right)_{(T_{hs}-T_a),p}, \quad (10)$$

$$T^{mhc}_{ORC1} = \frac{q_{c1}+q_{l1}}{\Delta s_{c1}}, \quad (11)$$

$$T^{mhc}_{ORC2} = \frac{q_{c2}+q_{l2}}{\Delta s_{c2}}, \quad (12)$$

where T^{mhs}_{ORC1} is the mean temperature of the condensation in the compared cycle, T^{mhs}_{ORC2} is the mean temperature of the condensation in the reference cycle, q_c is the heat of working fluid cooling, q_l is the heat of the working fluid liquefaction, Δs_c is the change in the working fluid specific entropy during cooling and liquefaction.

The comparison of two different working fluids and two different ORCs using the ψ_{TMC} parameter gives the opportunity of selecting the working fluid that will reject heat at the lowest possible heat rejection temperature $T_c \approx T_a$, i.e., the working fluid that will allow to lower the temperature of the heat sink and, therefore, increase the ORC system efficiency.

The process of the working fluid condensation is composed of the consecutive processes of cooling and liquefying. Rating parameters can also be defined for these sequential processes.

The amount of heat needed to be rejected from the working fluid to heat sink, to cool the vapor of the working fluid in the range of vapor quality of $0 < x < 1$ (see line between points 2 and c in Figure 1), and to liquefy the working fluid vapor (see line between points c and 3 in Figure 1) at a fixed heat sink temperature (T_c) may vary for different working fluids.

The rating parameter for cooling of the working fluid vapor can be defined by the equation

$$\psi_{C} = \left(\frac{q_{C1}}{q_{C2}}\right)_{(T_{hs}-T_a),p} \quad (13)$$

where q_{s1} is the amount of heat that has to be rejected from the working fluid to the heat sink to cool the working fluid in the compared cycle, q_{s2} is the amount of heat that has to be rejected from the working fluid to the heat sink to cool the working fluid in the reference cycle.

The rating parameter for liquefying the working fluid vapor can be defined by the equation

$$\psi_{L} = \left(\frac{q_{L1}}{q_{L2}}\right)_{(T_{hs}-T_a),p} \quad (14)$$

where q_{L1} is the amount of heat that has to be rejected from the working fluid to heat sink to liquefy the working fluid in the compared cycle, q_{L2} is the amount of heat that has to be rejected from the working fluid to the heat sink to liquefy the working fluid in the reference cycle.

The comparison of the different working fluids and different ORCs using the ψ_{TMC}, ψ_C, and ψ_L rating parameters gives the opportunity of selecting the working fluid that will limit the amount of heat needed to be rejected from the working fluid to heat sink to cool and liquefy the working fluid vapor and thus to lower the heat transfer area and dimensions of the applied cooler and condenser.

3.4. The Influence of the Pressure and Volumetric Expansion Ratio

The pressure and volumetric expansion ratio are important rating parameters influencing the pressure range in the cycle and operating conditions of the applied expander.

Pressure expansion ratio is given by the equation

$$\sigma_P = \left(\frac{p_1}{p_2}\right)_{(T_{hs}-T_a),p} \tag{15}$$

where p_1 is the working fluid pressure at the inlet to the expander, p_2 is the working fluid pressure at the outlet from the expander.

The volumetric expansion ratio is given by the equation

$$\sigma_{VEX} = \left(\frac{v_1}{v_2}\right)_{(T_{hs}-T_a),p} \tag{16}$$

where v_1 is the specific volume of the working fluid at the inlet to the expander, v_2 is the specific volume of the working fluid at the outlet from the expander.

The pressure expansion ratio σ_p describes the variation of the working fluid pressure during the expansion in the expander. This parameter can be applied for comparative analyses of the working fluids in terms of decreasing the pressure difference in the cycle and, in this way, simplifying the technical design of the ORC system (i.e., a lower pressure difference in the cycle means simpler sealing or thinner tubes can be applied). This parameter can also be used for selecting the volumetric expander in an ORC system using a specific working fluid and specific heat source and heat sink.

The volumetric expansion ratio σ_v describes the variation of the working fluid specific volume during the expansion in the expander. This parameter can be applied for comparative analyses of the working fluids in terms of increasing the specific volume of the working fluid during the expansion process. It may also be applied for selecting the volumetric expander in an ORC system using specified working fluid and a specified heat source and heat sink.

3.5. The Influence of the Volumetric Expandability of the Working Fluid during the Evaporation

The volumetric expandability of the working fluid during evaporation is another important parameter that has an influence on the selection of the volumetric expander in an ORC system. Volumetric expanders are usually characterized by a specified vapor consumption, which depends on the geometrical dimensions of the expander and rotational speed of the rotor (or shaft), i.e., the cycle frequency.

Volumetric expandability can be described by the equation

$$\sigma_V = \left(\frac{v_1}{v_4}\right)_{(T_{hs}-T_a),p} \tag{17}$$

where v_1 is the specific volume of the working fluid at the outlet of the evaporator, v_4 is the specific volume of the working fluid at the inlet to the evaporator.

The comparison of the different working fluids and different ORC cycles using the σ_V parameter gives the opportunity of selecting the working fluid for a specific volumetric expander or selecting the volumetric expander in a specified ORC system working with specified working fluid. This parameter can be useful for comparative analysis of working fluids in terms of the increase of the specific volume of the working fluid, i.e., the amount of vapor that can be obtained via evaporation of 1 kg of liquid working fluid.

In addition to the above-mentioned rating parameters, it is possible to define others related to, for example, critical parameters of the working fluids or their thermal properties, such as thermal conductivity, kinematic viscosity, thermal capacity, and others. However, these parameters will not be analyzed in the present study.

The above-defined rating parameters can be treated together as the working fluid selection method and applied for comparative selection of the working fluid for the ORC system, if the thermal characteristics of the heat source and the heat sink (namely the heat source and the heat sink temperatures) are known. The selection is based on the obtained maximum or minimum values of the rating parameters. Moreover, based on the comparison of the calculated parameters and the values of the expansion ratios, as well as the values of specific expansion work, it is possible to select a suitable volumetric expander for an ORC system.

The weight and importance of the rating parameters are always connected with the application of the ORC system. For example, if an ORC system is applied for cooling the machines and extracting heat from cooling mediums (e.g., in the case of internal combustion engines), heat removal from the cooling medium is the priority. Therefore, the most important rating parameter, in this case, will be the thermal capacity parameter (ψ_{TC}), and the maximum value of this parameter should be treated as the priority during the selection of the expander and working fluid.

In the following paragraph, a comparative analysis of selected working fluids using the above-described method is presented.

4. Comparative Analysis of the Working Fluids

4.1. The Analysis of the Influence of the Heat Source Characteristic and Histogram on the Pressure Expansion Ratio and the Obtained Isentropic Expansion Work

The analysis of the influence of the heat source characteristic and histogram on the pressure expansion ratio and the obtained isentropic expansion work was performed for histograms presented in Figure 5a–d. For the sake of clarity of the presented results, four working fluids (i.e., R123, R124, R134a, and R245fa) were selected for this comparison from the working fluids listed in Table 2. Calculations were performed for the following assumptions:

- Heat source temperature varies in the range of t_{hs} = 40–150 °C,
- Specific heat, which is supplied to the ORC system from the heat source, is calculated using the relation $q^+_{ORC} = h_1 - h_4$,
- Heat sink temperature is equal to t_c = 20 °C,
- Theoretical efficiency of the ORC cycle is calculated using the relation $\eta_{ORC} = l_{EXS}/q^+_{ORC} = \frac{h_1-h_{2s}}{h_1-h_4}$,
- Specific isentropic expansion work is calculated using the relation $l_{EXS} = q^+_{ORC}/\eta_{ORC}$,
- Heat source histograms have a course such as that presented in Figure 5a–d.

It should be reminded here again that each type of the volumetric expander features a specific highest value and optimum range of pressure expansion ratio σ_P (this parameter is also dependent on the design of the machine, i.e., its geometrical dimensions) for which the expander reaches the highest efficiency and power. The operational parameters of the expander should be kept in the range enabling the exploitation of the machine within the optimum range of the pressure expansion ratio. This issue was treated in more detail in [3,4]. For example, in the case of a multi-vane expander, it was reported in [57,58] that the highest internal efficiency is achieved for the value of $\sigma_P \approx 5$, and further increasing of the pressure expansion ratio has no effect on the expanders' internal efficiency. Examples of the influence of the pressure expansion ratio and rotational speed of the expander shaft on the expander power output are presented in Figure 7. As can be seen from this Figure, each curve has a visible maximum within a different range of rotational speed and pressure expansion ratio. Therefore, in order to keep the proper expander operating conditions and high values of the efficiency and power output, the machine operational parameters (i.e., rotational speed, pressure expansion ratio) should be regulated in a way that will enable to keep their values in a range located near the curve maximum.

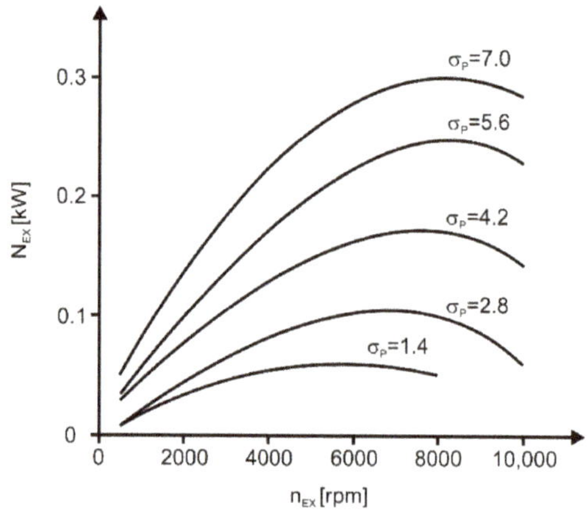

Figure 7. The influence of the expansion ratio and rotational speed of the expander shaft on the expander power output.

As reported in Table 1, each volumetric expander features the maximum value of the pressure expansion ratio (reaching $\sigma_p = 200$ for piston expanders). However, to the knowledge of the author, piston expanders reaching such a high pressure expansion ratio are not yet successfully applied in ORC systems. Therefore, in the following considerations, the obtained calculation results are limited to the value of $\sigma_p = 15$, which is the typical value of the highest pressure expansion ratio obtained experimentally in volumetric expanders that are successfully applied in commercial and research ORC units (i.e., screw, scroll, and multi-vane expanders). Figure 8 visualizes the variation of the isentropic expansion work for different characteristics of the heat source (i.e., histograms).

Figure 8a is valid for the histogram presented in Figure 5a. In the analyzed range of the pressure expansion ratio, the highest values of the isentropic expansion work are obtained for R123 (with the maximum value obtained for a pressure expansion ratio of $\sigma_p \approx 6.5$). For R134a, R124, and R245fa, the highest values of the isentropic expansion work were obtained for the pressure expansion ratio of $\sigma_p \approx 3.5$, $\sigma_p \approx 4.8$, and $\sigma_p \approx 15$, correspondingly. Similar results (see Figure 8b) were obtained for the heat source histogram presented in Figure 5b.

Different shapes of the characteristics (see Figure 8c) were obtained for the heat source histogram presented in Figure 5c. The highest value of isentropic expansion work was obtained for R123 and the pressure expansion ratio of $\sigma_p \approx 15$. For R134a, R124, and R245fa, the highest values of the isentropic expansion work were obtained for the pressure expansion ratio of $\sigma_p \approx 5$, $\sigma_p \approx 10$, and $\sigma_p \approx 15$, respectively.

Figure 8d visualizes the characteristics obtained for the heat source characteristics presented in Figure 5d. The highest values of isentropic expansion work are obtained for the pressure expansion ratio of $\sigma_p \approx 9$, $\sigma_p \approx 15$, $\sigma_p \approx 4$, and $\sigma_p \approx 6$ for R123, R245fa, R134a, and R124, respectively.

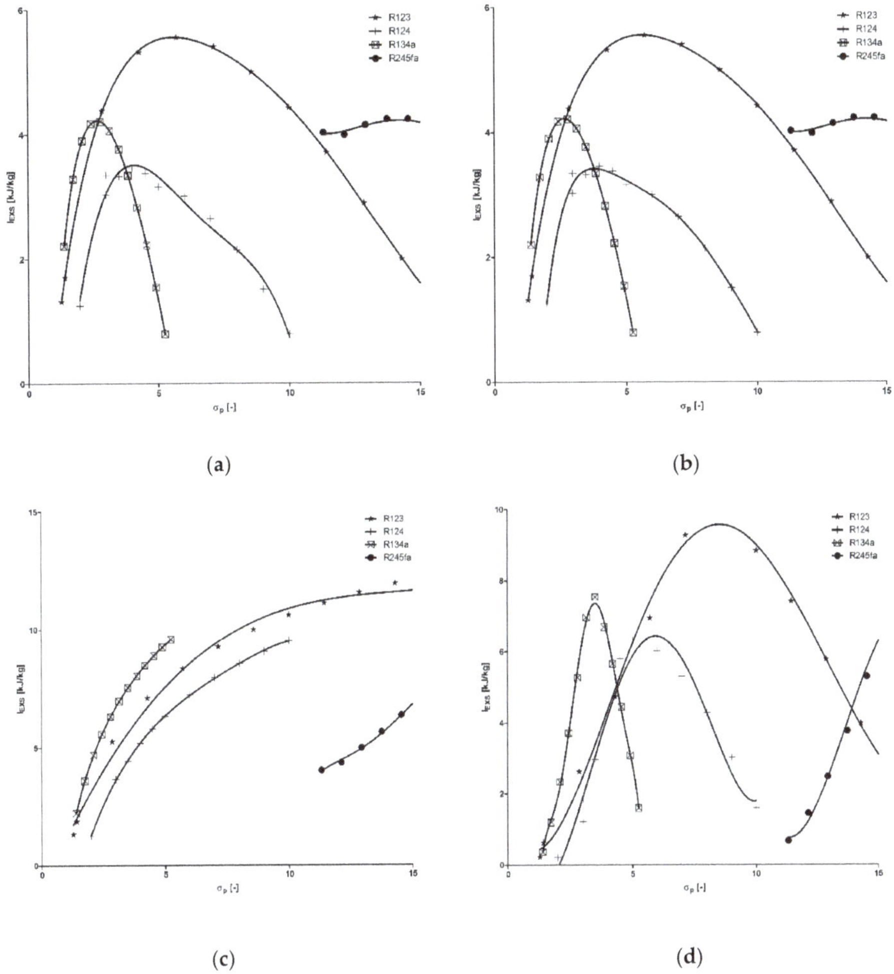

Figure 8. The variation of isentropic expansion work vs. pressure expansion ratio: (**a**) graph valid for the heat source histogram presented in Figure 5a; (**b**) graph valid for the heat source histogram presented in Figure 5b; (**c**) graph valid for heat source histogram presented in Figure 5c; (**d**) graph valid for the heat source histogram presented in Figure 5d.

By a comparative analysis of the results, which are visualized in Figure 8, the working fluid can be selected for the ORC system employing a specified volumetric expander, or a volumetric expander can be selected for the ORC system utilizing a specified working fluid. For example, as was previously stated, the optimum pressure expansion ratio ranges between four and five for multi-vane expanders that are applied in ORCs.

Therefore, if the highest value of the isentropic expansion work would be considered as the main working fluid selection criterion and the multi-vane expander would be considered as the prime mover in the ORC system, R123 (in the case of heat source histograms presented in Figure 5a,b) and R134a (in the case of heat source histograms presented in Figure 5c,d) are the working fluids in which the application should potentially result in the highest expander power.

4.2. The Comparative Analysis of the Working Fluids Using Rating Parameters

In the following part of this section, the comparative analysis of the working fluids using the rating parameters described in Sections 3.2–3.5 is presented. The analysis was performed for working fluids listed in Table 2 according to the process layout visualized in Figure 6. The values of thermal properties of the working fluids (i.e., specific enthalpy, specific entropy, etc.) were calculated using the CoolProp software.

The following assumptions were accepted for calculations:

- Cycle is evaporative, and superheating is not applied,
- Temperature of the heat source ranges between 40 °C and 150 °C,
- Expander internal efficiency is equal to $\eta_i = 0.7$, which is a typical mean value of volumetric expander internal efficiency [29],
- Temperature of the heat sink is equal to $t_c = 20$ °C,
- The reference substance is R123, which was selected based on the positive results of experiments on this working fluid, which were conducted by the author using the test-stands described in [29,30],
- ORC system is operating as the power plant (i.e., there is no heat generated for central heating),
- Heat losses in the ORC system are negligible.

The earlier defined rating parameters (i.e., ψ_{TC}, ψ_{TME}, ψ_{PH}, ψ_V, ψ_{TMC}, ψ_C, ψ_L) together with the pressure and volumetric expansion ratio (σ_P, σ_{VEX}), volumetric expandability (σ_V), efficiency of the cycle (η_{ORC}), and expander specific work (l_{EX}) were calculated for each of the above-mentioned working fluids. ψ_{SH} was not calculated because superheating is not considered in the present study. The results of these calculations are visualized in Figures 9–19. For the same reason as in the case of Figures 8–19 are limited to the value of $\sigma_P = 15$.

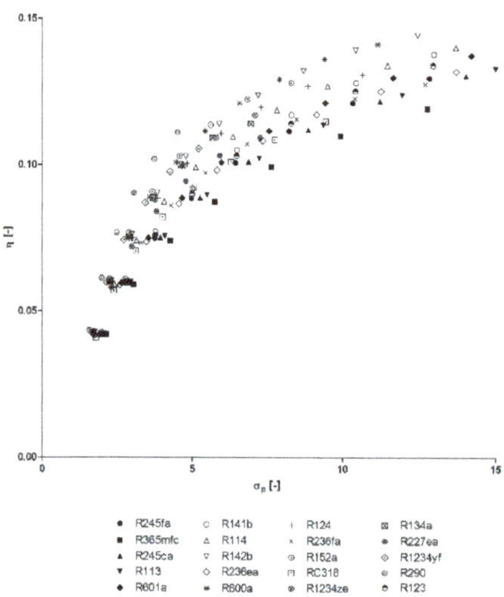

(a)

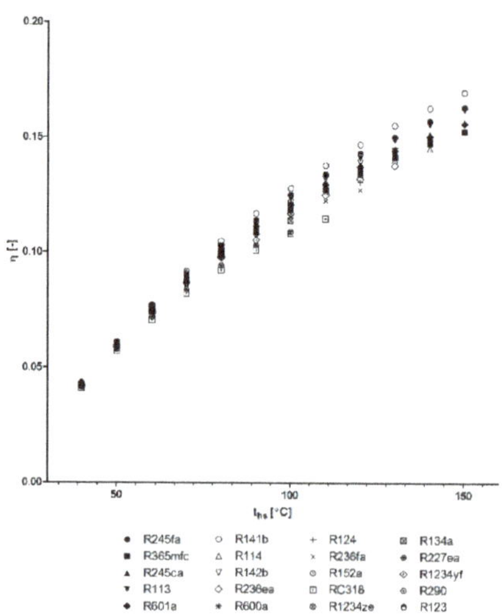

(b)

Figure 9. The variation of ORC system efficiency vs. pressure expansion ratio and heat source temperature: (**a**) $\eta = f(\sigma_P)$; (**b**) $\eta = f(t_{hs})$.

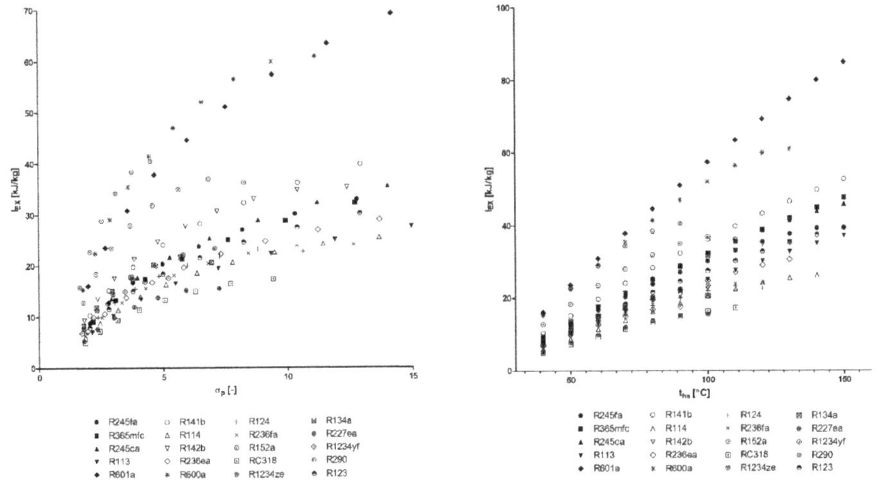

(a) (b)

Figure 10. The variation of isentropic expansion work vs. pressure expansion ratio and heat source temperature: (**a**) $l_{ex} = f(\sigma_P)$; (**b**) $l_{ex} = f(t_{hs})$.

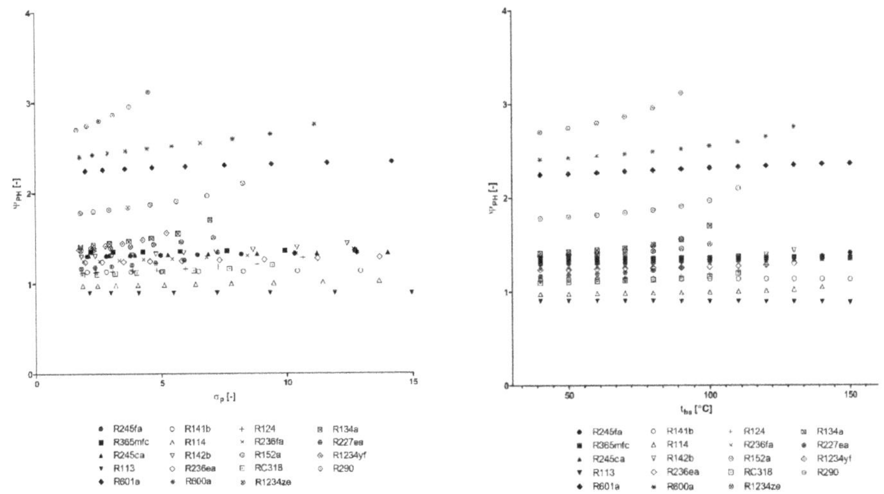

(a) (b)

Figure 11. The variation of the ψ_{PH} parameter vs. the pressure expansion ratio and heat source temperature: (**a**) $\psi_{PH} = f(\sigma_P)$; (**b**) $\psi_{PH} = f(t_{hs})$.

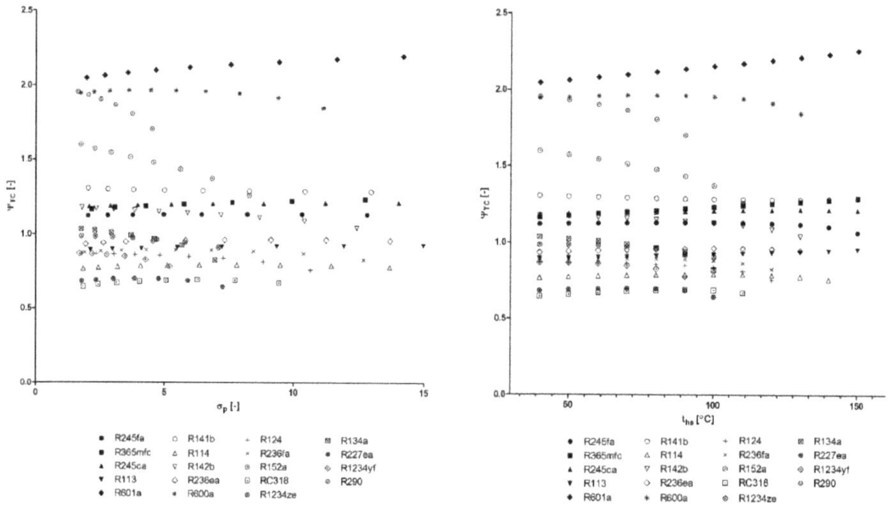

(a) (b)

Figure 12. The variation of the ψ_{TC} parameter vs. the pressure expansion ratio and heat source temperature: (**a**) $\psi_{TC} = f(\sigma_P)$; (**b**) $\psi_{TC} = f(t_{hs})$.

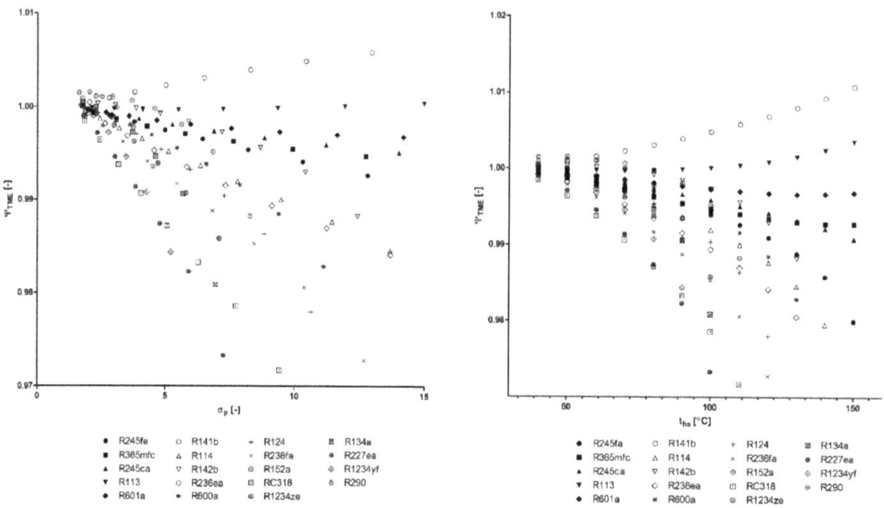

(a) (b)

Figure 13. The variation of ψ_{TME} parameter vs. pressure expansion ratio and heat source temperature: (**a**) $\psi_{TME} = f(\sigma_P)$; (**b**) $\psi_{TME} = f(t_{hs})$.

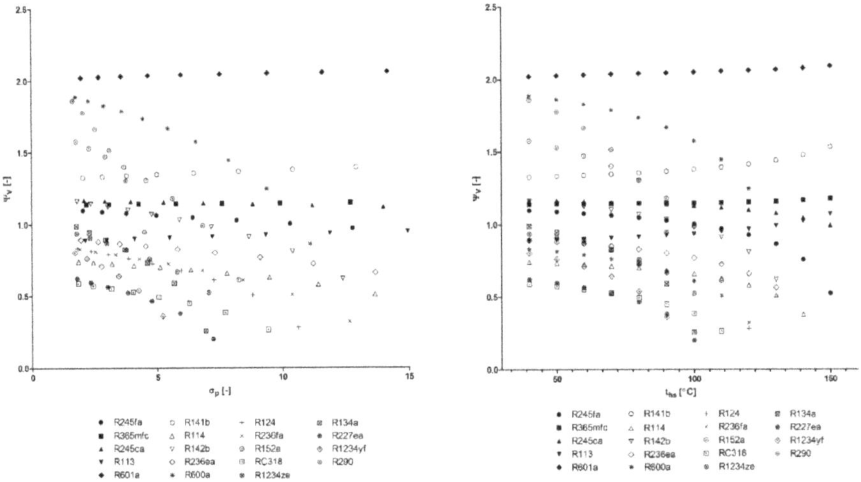

(a) **(b)**

Figure 14. The variation of the ψ_V parameter vs. the pressure expansion ratio and heat source temperature: (**a**) $\psi_V = f(\sigma_P)$; (**b**) $\psi_{TME} = f(t_{hs})$.

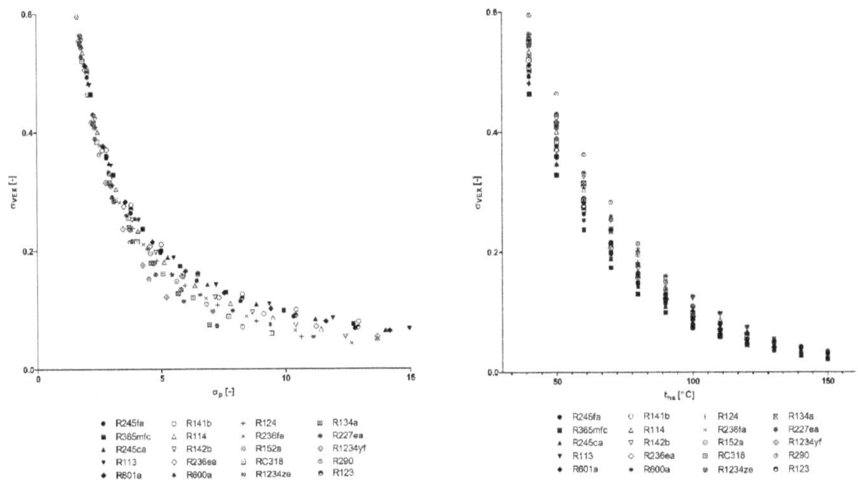

(a) **(b)**

Figure 15. The variation of the σ_{VEX} parameter vs. pressure expansion ratio and heat source temperature: (**a**) $\sigma_{VEX} = f(\sigma_P)$; (**b**) $\sigma_{VEX} = f(t_{hs})$.

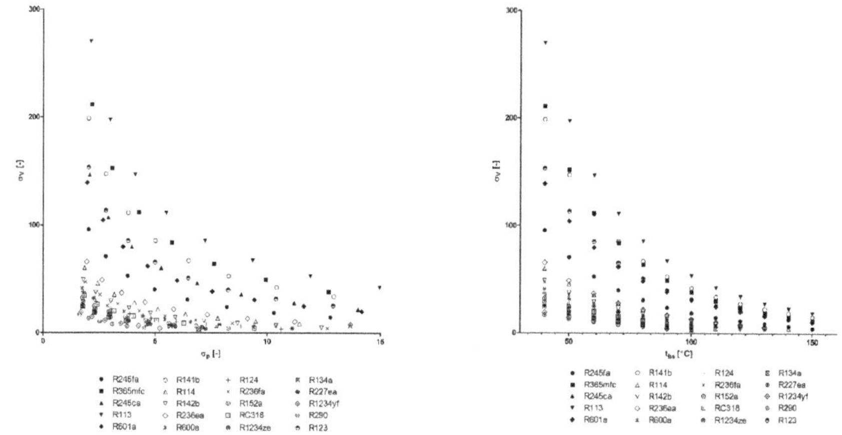

(a) (b)

Figure 16. The variation of the σ_V parameter vs. the pressure expansion ratio and heat source temperature: (**a**) $\sigma_V = f(\sigma_P)$; (**b**) $\sigma_V = f(t_{hs})$.

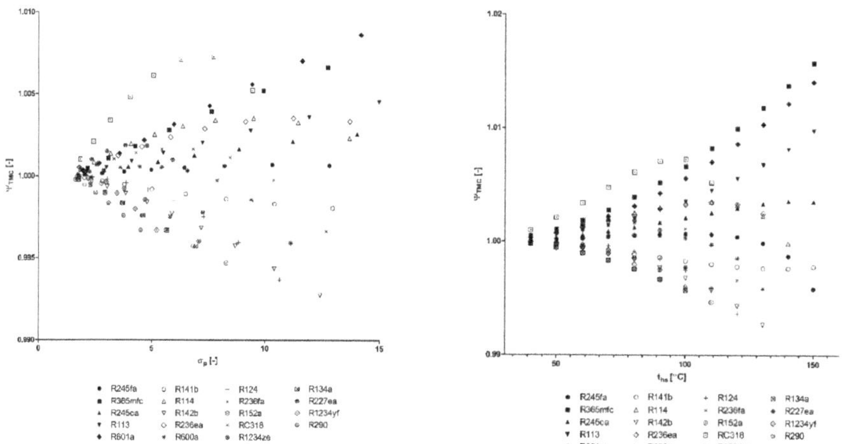

(a) (b)

Figure 17. The variation of the σ_V parameter vs. the pressure expansion ratio and heat source temperature: (**a**) $\sigma_{TMC} = f(\sigma_P)$; (**b**) $\sigma_{TMC} = f(t_{hs})$.

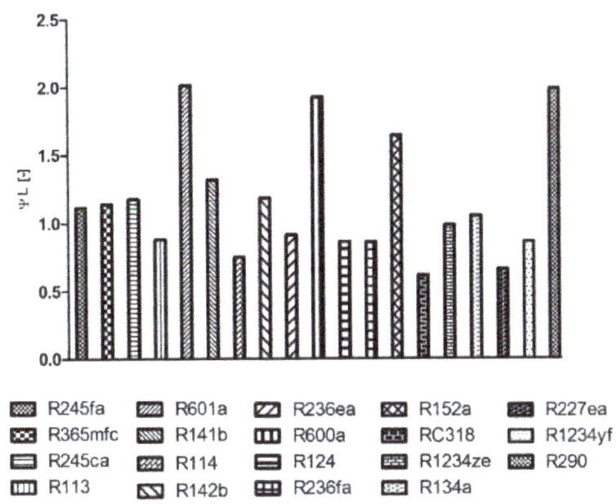

Figure 18. The variation of the ψ_L rating parameter (graph is valid for the heat sink temperature of $t_c = 20$ °C).

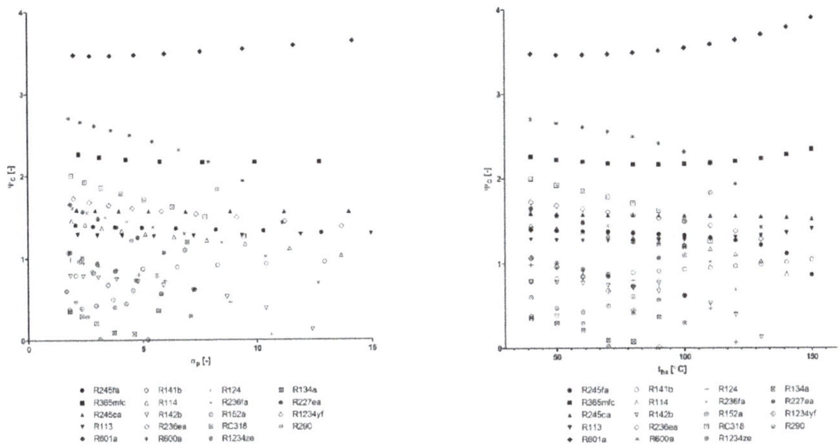

(**a**) (**b**)

Figure 19. The variation of the ψ_C parameter vs. the pressure expansion ratio and heat source temperature: (**a**) $\psi_C = f(\sigma_P)$; (**b**) $\psi_C = f(t_{hs})$.

Figure 9a shows the variation of ORC system efficiency vs. pressure expansion ratio. The highest efficiency (ca. 14%) was achieved for R142b and a pressure expansion ratio of $\sigma_p \approx 12.5$. Lower (but higher than in the case of R123) values of ORC system efficiency were obtained for R290, R600a, R152a, R124, R114, R1234ze, R1234yf, R227ea.

For the other working fluids, the obtained ORC system efficiency is lower than in the case of R123. The variation of ORC system efficiency vs. heat source temperature, which is presented in Figure 9b, shows that the maximum value of the ORC system efficiency is obtained for R141b in the considered range of heat source temperature. For example, if a multi-vane machine is selected as the expander, thence within the range of optimum pressure expansion ratio of these machines, the highest efficiency of the ORC system (ranging between 10% and 11%) was achieved for R290.

Figure 10a visualizes the variation of specific expansion work (l_{ex}) vs. the pressure expansion ratio. The highest value (ca. 70 kJ/kg) of specific expansion work was obtained for R601a and pressure expansion ratio $\sigma_p \approx 15$. Lower (but higher than in the case of R123) values of expansion work were obtained for R290, R600a, R142b, R141b, R245fa, R245ca, R134a, R1234yf, R365mfc, and R1234ze, which means that at the same working fluid mass flow rate through the expander, a higher power of the expander can be obtained using these working fluids compared to R123. For other working fluids, obtained values of isentropic expansion work were lower than for R123. The variation of expansion work vs. temperature of the heat source, which is presented in Figure 10b, shows that the highest value of specific expansion work is achieved for R601a in the considered range of heat source temperature. For example, if a multi-vane machine is selected as the expander, then within the range of the optimum pressure expansion ratio of these machines, the highest value of isentropic expansion work ranges between 35 and 40 kJ/kg and is obtained for R290 and R601a, correspondingly. Figure 11 shows the variation of the ψ_{PH} rating parameter vs. the pressure expansion ratio (see Figure 11a) and temperature of the heat source (see Figure 11b). The lowest values of this parameter ($\psi_{PH} < 1$) were achieved for R113 and R114 in the considered range of the pressure expansion ratio and heat source temperature.

Therefore, if the main criterion of the working fluid selection would be limiting the amount of heat required to preheat the working fluid (see process 4-a in Figure 1) and, therefore, to limit the heat transfer area and dimensions of the preheater, these working fluids are promising. For some working fluids analyzed (R600a, R601a, and R290), the amount of heat required to preheat the working fluid is even two or three times bigger than in the case of R123. For example, if the main application purpose of the ORC system is to exploit the heat source as much as possible (e.g., in the case of ORC application for waste heat recovery from piston engine cooling systems, etc.), these working fluids would be promising in that case.

Figure 12 shows the variation of the ψ_{TC} rating parameter vs. the pressure expansion ratio (see Figure 12a) and temperature of the heat source (see Figure 12b). The lowest values of this parameter ($\psi_{TC} < 1$) were achieved for RC318 and R227ea in the considered range of the pressure expansion ratio and heat source temperature.

Thus, if the main criterion of the working fluid selection would be limiting the amount of heat required for the working fluid phase transition (liquid–gas) to proceed (see process 4-b in Figure 1), these working fluids are promising in that case. For some working fluids analyzed (R600a, R601a, and R290), the amount of heat required to preheat the working fluid is even two times bigger than in the case of R123. For example, if the main application purpose of the ORC system is to exploit the heat source as much as possible (e.g., in the case of ORC application for waste heat recovery from piston engine cooling systems, etc.) these working fluids would be promising in that case.

Figure 13 shows the variation of the ψ_{TME} rating parameter vs. the pressure expansion ratio (see Figure 13a) and temperature of the heat source (see Figure 13b). The lowest values of this parameter ($\psi_{TME} < 1$) were achieved for RC318 and R227ea in the considered range of the pressure expansion ratio and heat source temperature. Therefore, if the main criterion of the working fluid selection would be lowering the heat source temperature (i.e., cooling the heat source) during the liquid–gas phase transition of the working fluid (see process 4-a in Figure 1), these working fluids are promising in that case. For some working fluids (R141b, R290, and R142b), the mean temperature of heat supply is higher than for R123 ($\psi_{TME} > 1$). If the purpose of the ORC system operation is heat extraction from the heat source at possibly high temperatures, these working fluids are promising in that case.

Figure 14 shows the variation of the ψ_V rating parameter vs. the pressure expansion ratio (see Figure 14a) and temperature of the heat source (see Figure 14b). The lowest values of this parameter ($\psi_V < 1$) were achieved for R1234yf and R227ea in the considered range of the pressure expansion ratio and heat source temperature. Thus, if the main criterion of working fluid selection would be lowering the heat amount that is required to evaporate the working fluid (see process a-b in Figure 1) and, therefore, to lower the heat transfer area and dimensions of the evaporator, these working fluids are promising in that case. For some working fluids (R600a, R601a, and R290), the amount of heat required to preheat the working fluid is even 1.5–2 times bigger than in the case of R123. For example, if the main application purpose of the ORC system is to exploit the heat source as much as possible (e.g., in the case of ORC application for waste heat recovery from piston engine cooling systems, etc.), these working fluids would be promising in that case.

Figure 15a visualizes the variation of the volumetric expansion ratio σ_{VEX} vs. the pressure expansion ratio. The volumetric expansion ratio is an important design parameter of the volumetric expander, which has an influence on the machine dimensions and volume of the working chambers in a way that determines the gas flow capacity of a machine.

Figure 15a visualizes the variation of the volumetric expansion ratio σ_{VEX} vs. the pressure expansion ratio. The volumetric expansion ratio is an important design parameter of the volumetric expander that has an influence on the machine dimensions and volume of the working chambers in a way that determines the gas flow capacity of a machine. Figure 16 shows the variation of the volumetric expandability σ_V vs. the pressure expansion ratio (see Figure 16a) and temperature of the heat source (see Figure 16b). In the case of volumetric machines, the volumetric flow of gas at the inlet to the expander is related to machines' gas flow capacity and should be kept at the required level in order to provide the continuity of machine operation and optimum operating conditions (rotational speed, efficiency, and power output). Therefore, this parameter should be maximized in the case of ORCs employing volumetric expanders. The highest values of this parameter (σ_V = 50–270) were achieved for R113, R365mfc, and R141b in the considered range of the pressure expansion ratio and heat source temperature.

Figure 17 shows the variation of the ψ_{TMC} rating parameter vs. the pressure expansion ratio (see Figure 17a) and temperature of the heat source (see Figure 17b).

The lowest values of this parameter ($\psi_{TMC} < 1$) were achieved for R290, R1234yf, R124, and R142b in the considered range of the pressure expansion ratio and heat source temperature. Thus, if the main criterion of the working fluid selection would be to lower the heat sink temperature, these working fluids are promising in that case. For some working fluids (RC318, R601a, and R365mfc), the average temperature of heat rejection is higher than for R123 ($\psi_{TMC} > 1$). Therefore, if the main criterion of working fluid selection would be to increase the temperature of the heat sink, these working fluids are promising in that case.

Figure 18 shows the variation of the ψ_L rating parameter for analyzed working fluids in the form of a bar graph (the graph is valid only for the assumed constant temperature of the heat sink, i.e., t_c = 20 °C). The lowest values of this parameter ($\psi_L < 1$) were achieved for RC318 and R227ea in the considered range of the ORC system operating parameters. Thus, if the main criterion of the working fluid selection would be limiting the amount of heat rejected from the gas during cooling (see process 2-c in Figure 1) and additionally limiting the heat transfer area and dimensions of the system cooler, these working fluids are promising in that case.

Figure 19 shows the variation of the ψ_C rating parameter vs. the pressure expansion ratio (see Figure 19a) and temperature of the heat source (see Figure 19b).

The lowest values of this parameter ($\psi_C < 1$) were achieved for R134a, R142b, R152a, R290, and R124 in the considered range of the pressure expansion ratio and heat source temperature. Thus, if the criterion of the working fluid selection would be limiting the amount of heat rejected by the working fluid during condensation (see process c-3 in Figure 1) and additionally limiting the heat transfer area and dimensions of the system condenser, these working fluids are promising in that case.

4.3. Summary

The first set of calculations was related to the working fluid selection, which is based on the variation of isentropic expansion work vs. the pressure expansion ratio σ_P and characteristic (i.e., histogram) of the heat source. If the highest value of isentropic expansion work would be considered as the main working fluid selection criterion, R123 seems to be a promising working fluid reaching the highest values of the expansion work in a wide range of the studied conditions and for all characteristics of the heat source. For other working fluids studied, the obtained values of isentropic expansion work are lower than in the case of R123. However, if additional working fluid selection criteria (which are related to the type of the applied volumetric expander and its design, such as pressure expansion ratio) are considered, then calculation results proved that other working fluids are also promising in the case of different types of volumetric expanders. Differences are related to the heat source characteristics. For example, if the highest value of the isentropic expansion work would be considered as the working fluid selection criterion and the multi-vane expander (featuring the optimum pressure expansion ratio of 4–5) as the prime mover of the ORC system, R123 (for the heat source histogram presented in Figure 5a,b) and R134a (for the heat source histogram presented in Figure 5c,d) are the working fluids in which the application should potentially result in the highest obtained expander power.

The second set of calculations was related to the comparative analysis of the working fluids based on the defined rating parameters (ψ_{TC}, ψ_{TME}, ψ_{PH}, ψ_{V}, ψ_{TMC}, ψ_{C}, ψ_{L}) together with pressure and the volumetric expansion ratio (σ_P, σ_{VEX}), volumetric expandability (σ_V), efficiency of the cycle (η_{ORC}), and specific expansion work (l_{EX}). R123 was selected as the reference substance based on the positive results of experiments on this working fluid, which were proceeded by the author using the test-stands and the results of the first part of calculations. The values of the rating parameters were calculated for 20 selected working fluids (see Table 2). Obtained results were then plotted in figures and compared. It was found that some of the analyzed working fluids have better thermal properties than the reference R123, i.e.,

- Highest values of the ORC system efficiency were obtained for R142b. However, it was found that the application of R290, R600a, R152a, R124, R114, R1234ze, R1234yf, R227ea can also result in a higher cycle efficiency than in the case of R123.
- Highest values of the specific expansion work were obtained for R601a. Moreover, it was found that the application of R290, R600a, R142b, R141b, R245fa, R245ca, R134a, R1234yf, R365mfc, and R1234ze can also result in higher values of the specific expansion work than R123, which means that at the same working fluid mass flow rate through the expander, a higher expander power can be obtained using these working fluids compared to R123.
- Application of RC318 and R227ea can limit the required amount of heat that is needed to be supplied to the working fluid from the heat source to perform the liquid–gas phase transition. The obtained results showed that by the application of R600a, R601a, and R290, the amount of required heat can be increased, which can be important if the application of the ORC system for cooling machines or devices is considered.
- By the application of RC318 and R227ea, it is possible to lower the temperature of the heat source (i.e., mean temperature of the heat supply) during the liquid–gas phase transition more rapidly compared to R123. It was also found that by the application of R141b, R290, and R142b, it is possible to extract heat from the heat source at higher temperatures compared to R123.
- By the application of R113 and R114, the amount of heat required to preheat the working fluid can be limited compared to R123, and therefore, the heat transfer area and dimensions of the preheater can be lowered. On the other hand, the application of R600a, R601a, and R290 can increase the amount of heat required to preheat the working fluid, which can be important if the application of the ORC system for cooling machines or devices is considered.
- By the application of R1234yf and R227ea, it is possible to limit the amount of heat required to evaporate the working fluid and, therefore, lower the heat transfer area and dimensions of the evaporator compared to R123. On the other hand, the application of R600a, R601a, and R290 can increase the amount of heat required to preheat the working fluid, which can be important when the application of an ORC system for cooling machines or devices is considered.
- The lowest values of the volumetric expansion ratio were obtained for R1234yf and R227ea, while the maximum values were achieved for R152a and R141b in the considered range of the pressure expansion ratio. In the considered range of the heat source temperature, the lowest values of this parameter were obtained for R365mfc, R113, and R245ca, while the maximum values

were achieved for R152a and R600a. The value of this parameter can be minimized (if a smaller expander is needed) or maximized (if a larger expander is needed).

- The highest values of volumetric expandability (high values of which are positive if a volumetric machine is considered as the ORC expander) were obtained for R113, R365mfc, and R141b.
- By the application of R290, R1234yf, R124, and R142b, it is possible to lower the temperature (i.e., mean temperature of the working fluid liquefaction) of the heat source during the gas–liquid phase transition more rapidly compared to R123. By the application of RC318, R601a, and R365mfc, it is possible to reject heat at higher temperatures compared to R123.
- By the application of RC318 and R227ea, it is possible to limit the amount of heat that is required to cool the gaseous working fluid and, therefore, lower the heat transfer area and dimensions of the cooler compared to R123.
- By the application of R134a, R142b, R152a, R290, and R124, it is possible to limit the amount of heat that is required to liquify the working fluid vapor and, therefore, lower the heat transfer area and dimensions of the condenser compared to R123.

5. Summary and Conclusions

This study presents the comparative method of the working fluid selection for an ORC system powered by a low temperature heat source and employing a volumetric expander. This method is based on a comparison of the selected thermal properties of working fluids by the application of the rating parameters describing the following thermal properties of the working fluids: the working fluid thermal capacity, mean temperature of evaporation, mean temperature of condensation, pressure and volumetric expansion ratio, volumetric expandability, as well as the heat of preheating, vaporization, superheating, cooling, and liquefaction. Moreover, isentropic expansion work was considered as the rating parameter. The presented method can be used for selecting the working fluid for the ORC system, which is supplied by a heat source and heat sink featuring a specified thermal power and temperature. Moreover, using this method, it is possible to select the volumetric expander for the ORC system based on the comparison of the calculated parameters, the pressure expansion ratio, and specific expansion work. The proposed method is universal and can be easily adopted for different working fluids, different assumptions, and ORCs employing turbines as the expanders. Therefore, using this method, it is possible to compare many different ORCs powered by different heat sources. The method can also be applied for comparing different ORCs to classical steam power plants. In order to illustrate this treatment, sets of calculations were performed in order to give an outlook for the selection of working fluids and a volumetric expander. Special attention was paid to a multi-vane expander. The calculation results showed the comparison of the working fluid candidates for different operating conditions of an ORC system. These results can be helpful for scientists and engineers dealing with ORCs and volumetric expanders for making quick estimations and comparisons of different ORCs.

Funding: This research received no external funding.

Acknowledgments: The author would like to thank Attila R. Imre for the invitation to publish this article.

Conflicts of Interest: The author declares no conflict of interest.

References

1. Bao, J.; Zhao, L. A review of working fluid and expander selections for organic Rankine cycle. *Renew. Sustain. Energy Rev.* **2013**, *24*, 325–342. [CrossRef]
2. Lai, N.A.; Wendland, M.; Fischer, J. Working fluid for high-temperature organic Rankine cycles. *Energy* **2011**, *36*, 199–211. [CrossRef]
3. Gnutek, Z.; Kolasiński, P. The application of rotary vane expanders in ORC systems—Thermodynamic description and experimental results. *J. Eng. Gas Turbines Power* **2013**, *135*, 61901. [CrossRef]
4. Gnutek, Z. *Sliding-Vane Rotary Machinery. Developing Selected Issues of One-Dimensional Theory*; Wrocław University of Technology Publishing: Wrocław, Poland, 1997.

5. Usman, M.; Pesyridis, A.; Cockerill, S.; Howard, T. Development and Testing of a Free Piston Linear Expander for Organic Rankine Cycle Based Waste Heat Recovery Application. In Proceedings of the 5th International Seminar on ORC Power Systems, Athens, Greece, 9–11 September 2019.
6. Kaczmarczyk, T.; Ihnatowicz, E.; Żywica, G.; Kiciński, J. Experimental investigation of the ORC system in a cogenerative domestic power plant with a scroll expanders. *Open Eng.* **2015**, *5*, 411–420. [CrossRef]
7. Gao, P.; Jiang, L.; Wang, L.W.; Wang, R.Z.; Song, F.P. Simulation and experiments on an ORC system with different scroll expanders based on energy and exergy analysis. *Appl. Therm. Eng.* **2015**, *75*, 880–888. [CrossRef]
8. Jradi, M.; Li, J.; Liu, H.; Riffat, S. Micro-scale ORC-based combined heat and power system using a novel scroll expander. *Int. J. Low Carbon Technol.* **2014**, *9*, 91–99. [CrossRef]
9. Hsu, S.-W.; Chiang, H.-W.D.; Yen, C.-W. Experimental Investigation of the Performance of a Hermetic Screw-Expander Organic Rankine Cycle. *Energies* **2014**, *7*, 6172–6185. [CrossRef]
10. Zhang, Y.-Q.; Wu, Y.-T.; Xia, G.-D.; Ma, C.-F.; Ji, W.-N.; Liu, S.-W.; Yang, K.; Yang, F.-B. Development and experimental study on organic Rankine cycle system with single-screw expander for waste heat recovery from exhaust of diesel engine. *Energy* **2014**, *77*, 499–508. [CrossRef]
11. Tang, H.; Wu, H.; Wang, X.; Xing, Z. Performance study of a twin-screw expander used in a geothermal organic Rankine cycle power generator. *Energy* **2015**, *90*, 631–642. [CrossRef]
12. Öhman, H.; Lundqvist, P. Screw expanders in ORC applications, review and a new perspective. In Proceedings of the 3rd International Seminar on ORC Power Systems, Brussels, Belgium, 12–14 October 2015.
13. Więckiewicz, H.; Cantek, L. *Volumetric Compressors—Atlas*, 2nd ed.; Gdańsk University of Technology Publishing: Gdańsk, Poland, 1985.
14. Kim, Y.M.; Shin, D.G.; Kim, C.G. Optimization of Design Pressure Ratio of Positive Displacement Expander for Vehicle Engine Waste Heat Recovery. *Energies* **2014**, *7*, 6105–6117. [CrossRef]
15. Dumont, O.; Parthoens, A.; Dickes, R.; Lemort, V. Experimental investigation and optimal performance assessment of four volumetric expanders (scroll, screw, piston and roots) tested in a small-scale organic Rankine cycle system. *Energy* **2018**, *165*, 1119–1127. [CrossRef]
16. Latz, G.; Erlandsson, O.; Skåre, T.; Contet, A.; Andersson, S.; Munch, K. Performance Analysis of a Reciprocating Piston Expander and a Plate Type Exhaust Gas Recirculation Boiler in a Water-Based Rankine Cycle for Heat Recovery from a Heavy Duty Diesel Engine. *Energies* **2016**, *9*, 495. [CrossRef]
17. Bianchi, M.; Branchini, L.; Casari, N.; De Pascale, A.; Melino, F.; Ottaviano, S.; Pinelli, M.; Spina, P.R.; Suman, A. Experimental analysis of a micro-ORC driven by piston expander for low-grade heat recovery. *Appl. Therm. Eng.* **2019**, *148*, 1278–1291. [CrossRef]
18. Wronski, J.; Imran, M.; Skovrup, M.J.; Haglind, F. Experimental and numerical analysis of a reciprocating piston expander with variable valve timing for small-scale organic Rankine cycle power systems. *Appl. Energy* **2019**, *247*, 403–416. [CrossRef]
19. Oudkerk, J.F.; Dickes, R.; Dumont, O.; Lemort, V. Experimental performance of a piston expander in a small-scale organic Rankine cycle. *Iop Conf. Ser. Mater. Sci. Eng.* **2015**, *90*, 12066. [CrossRef]
20. Glavatskaya, Y.; Podevin, P.; Lemort, V.; Shonda, O.; Descombes, G. Reciprocating Expander for an Exhaust Heat Recovery Rankine Cycle for a Passenger Car Application. *Energies* **2012**, *5*, 1751–1765. [CrossRef]
21. Li, G.; Zhang, H.; Yang, F.; Song, S.; Chang, Y.; Yu, F.; Wang, J.; Yao, B. Preliminary Development of a Free Piston Expander–Linear Generator for Small-Scale Organic Rankine Cycle (ORC) Waste Heat Recovery System. *Energies* **2016**, *9*, 300. [CrossRef]
22. Song, P.; Wei, M.; Liu, Z.; Zhao, B. Effects of suction port arrangements on a scroll expander for a small scale ORC system based on CFD approach. *Appl. Energy* **2015**, *150*, 274–285. [CrossRef]
23. Emhardt, S.; Tian, G.; Chew, J. A review of scroll expander geometries and their performance. *Appl. Therm. Eng.* **2018**, *141*, 1020–1034. [CrossRef]
24. Kim, D.; Chung, H.J.; Jeon, Y.; Jang, D.S.; Kim, Y. Optimization of the injection-port geometries of a vapor injection scroll compressor based on SCOP under various climatic conditions. *Energy* **2017**, *135*, 442–454. [CrossRef]
25. Li, J.; Yang, F.; Zhang, H.; Wu, Z.; Tian, Y.; Hou, X.; Xu, J.; Ren, J. Comparative analysis of different valve timing control methods for single-piston free piston expander-linear generator via an orthogonal experimental design. *Energy* **2020**, 116966. [CrossRef]

26. Read, M.; Smith, I.K.; Stosic, N. Optimisation of Screw Expanders for Power Recovery from Low Grade Heat Sources. *Energy Technol. Policy* **2014**, *1*, 131–142. [CrossRef]
27. Smith, I.K.; Stosic, N.; Kovacevic, A. *Power Recovery from Low Grade Heat by Means of Screw Expanders*; Elsevier: Cambridge, UK, 2014.
28. Lemmon, E.W.; Bell, I.H.; Huber, M.L.; McLinden, M.O. *NIST Standard Reference Database 23: Reference Fluid Thermodynamic and Transport Properties-REFPROP, Version 10.0*; National Institute of Standards and Technology, Standard Reference Data Program: Gaithersburg, MD, USA, 2018.
29. Bell, I.H.; Wronski, J.; Quoilin, S.; Lemort, V. Pure and Pseudo-pure Fluid Thermophysical Property Evaluation and the Open-Source Thermophysical Property Library CoolProp. *Ind. Eng. Chem. Res.* **2014**, *53*, 2498–2508. [CrossRef] [PubMed]
30. Györke, G.; Deiters, U.K.; Groniewsky, A.; Lassu, I.; Imre, A.R. Novel classification of pure working fluids for Organic Rankine Cycle. *Energy* **2018**, *145*, 288–300. [CrossRef]
31. Marx, V.; Pruss, A.; Wagner, W. *Neue Zustandsgleichungen fuer R 12, R 22, R 11 und R 113. Beschreibung des thermodynamishchen Zustandsverhaltens bei Temperaturen bis 525 K und Druecken bis 200 MPa*; Series 19 (Waermetechnik/Kaeltetechnik), 57; VDI Verlag: Duesseldorf, Germany, 1992.
32. Platzer, B.; Polt, A.; Maurer, G. *Thermophysical Properties of Refrigerants*; Springer: Berlin/Heidelberg, Germany, 1990.
33. Younglove, B.A.; McLinden, M.O. An International Standard Equation of State for the Thermodynamic Properties of Refrigerant 123 (2,2-Dichloro-1,1,1-trifluoroethane). *J. Phys. Chem. Ref. Data* **1994**, *23*, 731–779. [CrossRef]
34. de Vries, B.; Tillner-Roth, R.; Baehr, H.D. Thermodynamic Properties of HCFC 124. In Proceedings of the 19th International Congress of Refrigeration, The Hague, The Netherlands, 20–25 August 1995.
35. McLinden, M.O.; Thol, M.; Lemmon, E.W. Thermodynamic Properties of trans-1,3,3,3-Tetrafluoropropene [R1234ze(E)]: Measurements of Density and Vapor Pressure and a Comprehensive Equation of State. In Proceedings of the International Refrigeration and Air Conditioning Conference at Purdue, Purdue University, West Lafayette, IN, USA, 12–15 July 2010.
36. Tillner-Roth, R.; Baehr, H.D. An international standard formulation of the thermodynamic properties of 1,1,1,2-tetrafluoroethane (HFC-134a) for temperatures from 170 K to 455 K at pressures up to 70 MPa. *J. Phys. Chem. Ref. Data* **1994**, *23*, 657–729. [CrossRef]
37. Outcalt, S.L.; McLinden, M.O. A modified Benedict-Webb-Rubin equation of state for the thermodynamic properties of R152a (1,1-difluoroethane). *J. Phys. Chem. Ref. Data* **1996**, *25*, 605–636. [CrossRef]
38. Lemmon, E.W.; Span, R. Thermodynamic Properties of R-227ea, R-365mfc, R-115, and R-13I1. *J. Chem. Eng. Data* **2015**, *60*, 3745–3758. [CrossRef]
39. Outcalt, S.L.; McLinden, M.O. *An Equation of State for the Thermodynamic Properties of R236fa*; NIST Report to Sponsor (U.S. Navy, David Taylor Model Basin) under Contract N61533-94-F-0152; NIST: Gaithersburg, MD, USA, 1995.
40. Defibuagh, D.R.; Gillis, K.A.; Moldover, M.R.; Schmidt, J.W.; Weber, L.A. Thermodynamic properties of CHF(2)-CF(2)-CH(2)F, 1,1,2,2,3-pentafluoropropane. *Int. J. Refrig.* **1996**, *19*, 285–294. [CrossRef]
41. Lemmon, E.W.; Span, R. Short Fundamental Equations of State for 20 Industrial Fluids. *J. Chem. Eng. Data* **2006**, *51*, 785–850. [CrossRef]
42. Defibaugh, D.R.; Gillis, K.A.; Moldover, M.R.; Schmidt, J.W.; Weber, L.A. Thermodynamic properties of CF3-CF-CHF2, 1,1,1,2,3,3-hexafluoropropane. *Fluid Phase Equilibria* **1996**, *122*, 131–155. [CrossRef]
43. Buecker, D.; Wagner, W. Reference Equations of State for the Thermodynamic Properties of Fluid Phase n-Butane and Isobutane. *J. Phys. Chem. Ref. Data* **2006**, *35*, 929–1019. [CrossRef]
44. Richter, M.; McLinden, M.O.; Lemmon, E.W. Thermodynamic Properties of 2,3,3,3-Tetrafluoroprop-1-ene (R1234yf): P-rho-T Measurements and an Equation of State. *J. Chem. Eng. Data* **2011**, *56*, 3254–3264. [CrossRef]
45. Lemmon, E.W.; McLinden, M.O.; Wagner, W. Thermodynamic Properties of Propane. III. A Reference Equation of State for Temperatures from the Melting Line to 650 K and Pressures up to 1000 MPa. *J. Chem. Eng. Data* **2009**, *54*, 3141–3180. [CrossRef]
46. Imre, A.R.; Kustán, R.; Groniewsky, A. Thermodynamic Selection of the Optimal Working Fluid for Organic Rankine Cycles. *Energies* **2019**, *12*, 2028. [CrossRef]

47. Zhang, X.; Zhang, Y.; Cao, M.; Wang, J.; Wu, Y.; Ma, C. Working Fluid Selection for Organic Rankine Cycle Using Single-Screw Expander. *Energies* **2019**, *12*, 3197. [CrossRef]
48. White, J.A.; Velasco, S. Approximating the Temperature–Entropy Saturation Curve of ORC Working Fluids from the Ideal Gas Isobaric Heat Capacity. *Energies* **2019**, *12*, 3266. [CrossRef]
49. Invernizzi, C.M.; Ayub, A.; Di Marcoberardino, G.; Iora, P. Pure and Hydrocarbon Binary Mixtures as Possible Alternatives Working Fluids to the Usual Organic Rankine Cycles Biomass Conversion Systems. *Energies* **2019**, *12*, 4140. [CrossRef]
50. Hung, T.C.; Wang, S.K.; Kuo, C.H.; Pei, B.S.; Tsai, K.F. A study of organic working fluids on system efficiency of an ORC using low-grade energy sources. *Energy* **2010**, *35*, 1403–1411. [CrossRef]
51. Wang, E.H.; Zhang, H.G.; Fan, B.Y.; Ouyang, M.G.; Zhao, Y.; Mu, Q.H. Study of working fluid selection of organic Rankine cycle (ORC) for engine waste heat recovery. *Energy* **2011**, *36*, 3406–3418. [CrossRef]
52. Setiawan, D.; Subrata, I.D.M.; Purwanto, Y.A.; Tambunan, A.H. Evaluation of Working Fluids for Organic Rankine Cycle Based on Exergy Analysis. *IOP Conf. Ser. Earth Environ. Sci.* **2018**, *147*, 12035. [CrossRef]
53. Siddiqi, M.A.; Atakan, B. Investigation of the Criteria for Fluid Selection in Rankine Cycles for Waste Heat Recovery. *Int. J. Thermodyn.* **2011**, *14*, 117–123.
54. Dai, X.; Shi, L.; Qian, W. Review of the Working Fluid Thermal Stability for Organic Rankine Cycles. *J. Therm. Sci.* **2019**, *14*, 597–607. [CrossRef]
55. Badr, O.; Hussein, M.; Probert, S.D.; O'Callaghan, P.W. Thermal Stabilities of Mixtures of Trichlorofluoroethane and Lubricating Fluids Contained in Copper Sealed Tubes. *Appl. Energy* **1984**, *16*, 41–52. [CrossRef]
56. Kolasiński, P. The method of the working fluid selection for organic Rankine cycle (ORC) system with volumetric expander. In Proceedings of the 3rd International Seminar on ORC Power Systems, Brussels, Belgium, 12–14 October 2015.
57. Kolasiński, P. The Influence of the Heat Source Temperature on the Multivane Expander Output Power in an Organic Rankine Cycle (ORC) System. *Energies* **2015**, *8*, 3351–3369. [CrossRef]
58. Kolasiński, P.; Błasiak, P.; Rak, J. Experimental and Numerical Analyses on the Rotary Vane Expander Operating Conditions in a Micro Organic Rankine Cycle System. *Energies* **2016**, *9*, 606. [CrossRef]

energies

Article

Zeotropic Mixture Selection for an Organic Rankine Cycle Using a Single Screw Expander

Xinxin Zhang [1,2,*], Yin Zhang [1,2], Zhenlei Li [1,2], Jingfu Wang [1,2], Yuting Wu [1,2] and Chongfang Ma [1,2]

1 MOE Key Laboratory of Enhanced Heat Transfer and Energy Conservation, College of Environmental and Energy Engineering, Beijing University of Technology, Beijing 100124, China; yinzhang@emails.bjut.edu.cn (Y.Z.); lizl@emails.bjut.edu.cn (Z.L.); jfwang@bjut.edu.cn (J.W.); wuyuting@bjut.edu.cn (Y.W.); machf@bjut.edu.cn (C.M.)

2 Beijing Key Laboratory of Heat Transfer and Energy Conversion, College of Environmental and Energy Engineering, Beijing University of Technology, Beijing 100124, China

* Correspondence: xinxinzhang@bjut.edu.cn; Tel.: +86-10-6739-1985

Received: 20 December 2019; Accepted: 21 February 2020; Published: 25 February 2020

Abstract: The organic Rankine cycle (ORC) is a popular and promising technology that has been widely studied and adopted in renewable and sustainable energy utilization and low-grade waste heat recovery. The use of zeotropic mixtures in ORC has been attracting more and more attention because of the possibility to match the temperature profile of the heat source by non-isothermal phase change, which reduces the irreversibility in the evaporator and the condenser. The selection of working fluid and expander is strongly interconnected. As a novel expander, a single screw expander was selected and used in this paper for efficient utilization of the wet zeotropic mixtures listed in REFPROP 9.1 in a low-temperature subcritical ORC system. Five indicators, namely net work, thermal efficiency, heat exchange load of condenser, temperature glide in evaporator, and temperature glide in condenser, were used to analyze the performance of an ORC system with wet and isentropic zeotropic mixtures as working fluids. The calculation and analysis results indicate that R441A with an expander outlet temperature of 320 K may be the suitable zeotropic mixture used for both open and close type heat source. R436B may be selected with an expander outlet temperature of 315 K. R432A may be selected with an expander outlet temperature from 295 K to 310 K.

Keywords: wet zeotropic mixture; single screw expander; organic Rankine cycle; R441A; R436B; R432A

1. Introduction

The characteristics of the world energy structure, which is dominated by fossil energy, have caused two long-term problems, namely, the depletion of fossil energy and the environmental pollution caused by the utilization of fossil energy. In order to solve these two problems, we have to take all measures that are technically feasible, economically reasonable, environmentally and socially acceptable to improve the utilization efficiency of energy resources. Among these measures, the organic Rankine cycle (ORC) is a popular and promising technology that has been widely studied and adopted in renewable and sustainable energy utilization and low-grade waste heat recovery. Working fluid and expander are two research hotpots of ORC. The selection of working fluid and expander is strongly interconnected. Water is used as the working fluid of steam Rankine cycle which is usually used to exploit and utilize a heat source higher than 450 °C. Turbine that has a high efficiency is used as the expander in steam Rankine cycle. Organic substance is used as the working fluid of the organic Rankine cycle which is usually used for the exploitation and utilization of medium-low grade heat source whose temperature is lower than 250 °C. Unlike lower molecular weight fluids like water, turbine design considerations in less than 100 kW output capacities results in lower efficiencies compared to heavier molecular weight organic fluids. The thermodynamic performance, working condition, impact on the environment, and economic feasibility of an ORC system are greatly determined by the characteristics of working fluid [1,2]. CFCs that were invented in 1930s and have a high ozone depletion potential (ODP) and a highest Global Warming Potential (GWP), HCFCs that were invented in 1950s and have a lower ODP and a high GWP, and HFCs that were invented in 1990s and have no ODP but a high GWP, were dominant organic working fluids before the ratification of the Montreal Protocol in 1987 and the ratification of the Kyoto Protocol in 1997. Changes are driven by regulations. Only those working fluids with zero ODP and a very low GWP can be used at present and in the future. Mixtures (blends) can meet this requirement. Basically, mixtures can be classified into two categories: azeotropes that have a constant boiling point and composition throughout distillation and zoetropes that boil across a range of temperatures at any given pressure. The use of zeotropic fluid mixtures in energy conversion systems has been widely studied for refrigeration plants and heat pumps in the last few decades [3]. The earliest research on zeotropic refrigerants can be traced back to the late 1980s [4] and early 1990s [5,6]. Radermacher proposed several solution circuits used to eliminate the inherent requirement of complete phase changes in the heat exchangers [4,5]. Weng experimentally examined the heat transfer performance of the zeotropic refrigerant blends of carbon tetrafluoride (R14) and dichlorotetrafluoroethane (R114) during evaporative flow under various conditions [6]. Nowadays, the use of zeotropic mixtures in power cycles has been attracting more and more attention because of the possibility to match the temperature profile of the heat source by non-isothermal phase change, which reduces the irreversibility in the evaporator and the condenser. If the irreversibility of the heat transfer process is reduced and given the low-grade heat source, the potential of harnessing useful work from the heat source is increased [7]. This advantage is very important and useful to ORC system. Therefore, a lot of research work has been conducted. Lecompte et al. examined the thermodynamic performance of a non-superheated subcritical ORC with seven zeotropic mixture pairs as working fluid. They found the evaporator accounts for the highest exergy loss and the matching with the condenser heat profiles results the best performance [8]. Zhao and Bao proposed a thermodynamic model which mainly includes Jacob number. They used the ratio of evaporation temperature to the condensation temperature to study the thermal efficiency, output work, and exergy efficiency of ORC system with ten zeotropic mixture pairs as working fluid. The significant influence of heat source inlet temperature on the best composition of zeotropic mixtures was found. Compared with pure fluids, the zeotropic mixture performance can be improved by greater temperature glide of the mixture [9]. Habka and Ajib conduct a performance analysis of zeotropic mixtures used in ORC systems for geothermal water utilization. They found R438A, R422A, and R22M are more efficient than the pure fluids and can enhance the power productivity and geothermal water utilization at the source's temperatures of 80, 100, and 120 °C, respectively [10]. Deethayat et al. proposed a dimensionless term named "Figure of Merit" (FOM) and studied the thermal performance of six zeotropic mixtures used in a low-temperature ORC system. They also developed an empirical correlation which fits very well with literature to estimate the cycle efficiency from the FOM for all working fluids at condensing temperatures of 25–40 °C and evaporating temperatures of 80–130 °C [11]. Miao et al. proposed a thermodynamic selection criterion of zeotropic mixtures based on the exergy analysis of the subcritical ORC. They found the match condition with the heat source should be firstly satisfied when selecting working fluids. The proper temperature glide in the condenser can further improve the cycle performance [12]. Battista et al. developed a comprehensive thermodynamic model of the ORC plant, considering both hot and cold source available on board vehicle. They evaluated the best candidates based on: thermodynamic performance, pressure levels, fluid hazard levels and GWP, critical temperature, and the temperature glide. They found R245fa is a fluid that obtains a large net power increase when used in mixtures with hydrocarbons, compared to pure fluid an optimized R245fa/benzene mixture, for instance, attains an 11% net power increase [13]. Cipollone et al. presented a thermodynamic analysis of a trilateral flash cycle (TFC) system using recent pure fluids and mixtures for low grade heat to power conversion applications. They found mixtures appear more suitable for rotary volumetric machines having lower built in volume ratios [14].

The results given in different research work indicate that zeotropic mixtures show a better thermodynamic performance at low heat source temperatures than high heat source temperatures [3]. According to thermodynamic knowledge, not much exergy is available in low-temperature heat source compared with high-temperature heat source. Therefore, in order to make full use of the exergy in low-temperature heat source, selection of proper zeotropic mixture as working fluid is very important. Same as pure working fluids, mixtures can be classified by three types of saturated vapor curve as shown on *T-s* diagram, namely dry, with a positive slope, wet, with a negative slope and isentropic, with a vertical slope. [15]. Miao et al. claimed that the 'wet' mixtures have relatively lower cycle performance compare to 'dry' and 'isentropic' ones [12]. However, their conclusion is based on using a turbine as an expander. In order to avoid the damage on turbine blades caused by liquid droplets, a vapor quality that is high enough must be ensured. Therefore, superheating apparatus is needed for wet fluids when turbo-type expander/turbine is used in ORC system. However, second law analysis showed that superheating organic fluids increases the irreversibility and decreases the second law efficiency [16]. Moreover, since the objective of the ORC focuses on the use of heat at low and medium temperatures, the overheating of the vapor, as in the traditional steam Rankine cycle, is not appropriate and is a waste of exergy in low-and-medium heat sources [17–19].

Zeotropic mixtures can be artificially defined and mixed into dry or isentropic working fluids that are suitable for ORC. However, if we screen predefined mixtures listed in REFPROP 9.1 [20], it can be seen that most zeotropic mixtures are wet ones. They are applicable in ORC and they do not need to be superheated in ORC if an expander which can tolerate wet expansion of wet zeotropic mixtures can be found. This indicates that the selection of working fluid and expander is strongly interconnected.

Expander is the critical device in an ORC system because it significantly affects thermodynamic performance of an ORC system and its cost ranks second in total system investment [21]. It includes turbo type and positive-displacement type. It has been mentioned that turbo type expander that are normally suitable for large scale ORC systems [19,22] but might not be favorable for small scale ORC units [23] and cannot be used as an expander for wet zeotropic mixture applications. If we screen positive-displacement expanders, such as rolling piston expander, scroll expander, and single screw expander, it can be found that the single screw expander is most suitable for wet zeotropic mixture application. It has the common advantages of positive-displacement expander, such as relatively high efficiency, high pressure ratio, low rotational speed, and tolerance of two-phase fluids [22]. Compared with rolling piston expander and scroll expander, single screw expander has many other advantages, such as balanced load of the screw, long service life, high volumetric efficiency, good performances in partial load, low leakage, low noise, low vibration, and simple configuration [23]. All these advantages have been attracting many researchers to carry out relevant study on design and application of single screw expander. The authors' research group has conducted a series of research works in the field of single screw expanders [23–31]. Normal and novel prototypes of single screw expanders have been experimentally studied [24,26,29]. The factors influencing the performance of a single screw expander have also been studied [26–28,30]. The applications of single screw expanders in ORC and refrigeration systems have also been introduced [23,25,31]. Brief descriptions and analyses of these research works can be found in our previous paper [32]. In a single screw structure, one screw can mesh with two or more starwheels. According to the shape, screw and starwheel can be divided into cylindrical (type C) and flat (type P). Therefore, these two types can be combined into four forms of single screw structure, namely PC type, PP type, CC type and CP type. These four forms are depicted in Figure 1. Because the first three forms are more difficult to process, the most commonly used single screw structure is the CP type. Figure 2 depicts the configuration of CP type single screw expander. Its working processes are depicted by Figure 3.

A demonstration project of ORC system using single screw expander with R123 as working fluid was established in Liulin (Shanxi Province, China). This project is used for waste heat recovery of the flue gas from a gas-fired internal combustion engine generator unit. The working condition parameters of the demonstration ORC system are listed in Table 1. The thermal efficiency of the ORC system is 8.81%. The output power of the demonstration project is 11 kW. Considering that R123 will be eliminated in the near future due to environmental factors, a selection of a pure working fluid for ORC using a single screw expander was conducted in our previous paper [21]. It is found that *cis*-butene may be the best candidate for working in a subcritical cycle. HFO working fluids are more suitable for working in near-critical cycles and HFO-1234ze(E) may be the best. In this paper, in order to efficiently use the wet zeotropic mixture in low-temperature subcritical ORC system, a single screw expander is selected and used in the system for analysis. Among all the predefined mixtures listed in REFPROP 9.1, four wet zeotropic mixtures and two isentropic zeotropic mixtures are selected to be the candidates for further calculation and analysis. On this basis, zeotropic mixture selection for ORC using single screw expander is conducted based on five indicators, namely net work, thermal efficiency, heat exchange load of condenser, temperature glide in evaporator, and temperature glide in condenser.

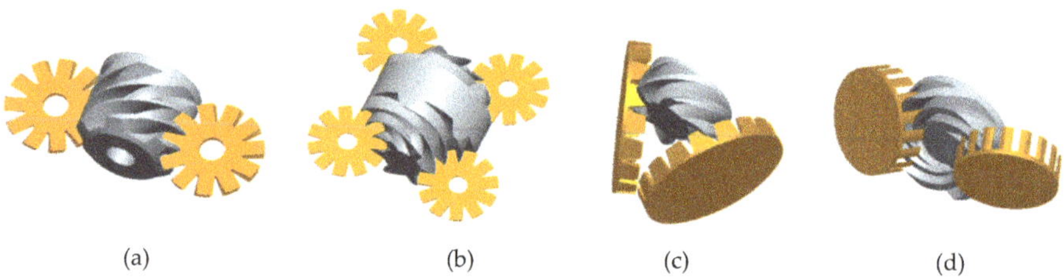

Figure 1. Four forms of single screw structure: (**a**) CP type; (**b**) CC type; (**c**) PP type; (**d**) PC type.

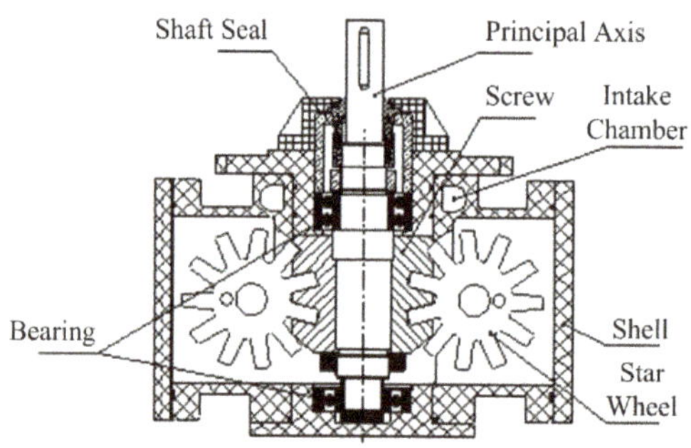

Figure 2. Configuration of CP type single screw expander.

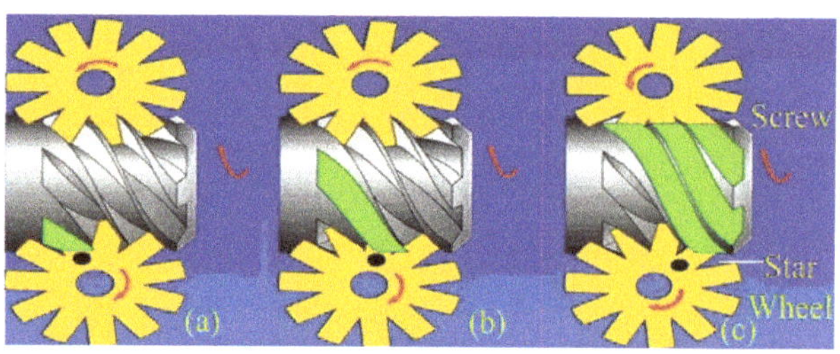

Figure 3. Working processes of CP type single screw expander (**a**) suction, (**b**) expansion, and (**c**) discharge.

Table 1. Working condition parameters of the demonstration ORC system

Working Condition Parameter.	Value
Heat source temperature (°C)	330–450
Mass flow rate of flue gas (heat source)	1700
Inlet pressure of expander (MPa)	1.6
Inlet temperature of expander (°C)	135
Working fluid flow(kg/h)	3100
Type of evaporator	shell-and-tube heat exchanger
Evaporation temperature (°C)	108–118
Type of condenser	tube-in-tube heat exchanger
Condensation temperature (°C)	30
Cooling water temperature (°C)	20

2. Basic Constraints and Preliminary Screening

Low-temperature heat sources can be divided into two categories: open type and closed type [33,34]. The inlet temperature and mass flow rate of an open-type heat source is known. The working mass of the heat source is directly discharged after being used. For closed type, the heat release is specific and the working mass of heat source is usually recycled after releasing heat. Therefore, different standards are used to measure these two types of heat source [34]. The maximum net work and the maximum thermal efficiency are used as the criteria for open type and closed type, respectively. Therefore, net work and thermal efficiency are adopted as the first two indicators for evaluating the performance of ORC using single screw expander with zeotropic mixture as working fluid. The third evaluation indicator is heat exchange load of condenser because it is the critical for calculating the cost of condenser that greatly influences the cost and economic performance of entire ORC system [21].

Zeotropic mixture performance can be improved by a greater temperature glide of the mixture [9]. In particular the temperature glide in the condenser can further improve the cycle performance [12]. Therefore, the fourth and fifth evaluation indicators are taken as temperature glide in evaporator and temperature glide in condenser.

Seventy nine (79) kinds of mixtures can be found in the REFPROP 9.1 software developed by the National Institute of Standards and Technology Laboratories (NIST, Gaithersburg, MD, USA) [20]. Among all these mixtures, 72 kinds of mixtures have identifying numbers given by American Society of Heating, Refrigerating and Air-Conditioning Engineers (ASHRAE). According to the latest ASHRAE standard [35], zeotropic blends shall be assigned an identifying number in the 400 series and azeotropes shall be assigned an identifying number in the 500 series. Therefore, 61 kinds of zeotropic mixtures with identifying numbers in the 400 series can be selected. Their basic properties are listed in Table 2.

From Table 2, it can be seen that HCFCs, such as R22, R124, and R142b, and HFCs, such as R32, R134a, R143a, R152a, and R125, are the components of zeotropic mixtures. HCFCs and HFCs have been banned due to their high GWP. Therefore, the zeotropic mixtures containing HCFCs and HFCs are screened out for further analysis. The remaining six zeotropic mixtures for further analysis, including four wet ones and two isentropic ones, and their critical temperatures are listed in Table 3. The components of the remaining zeotropic mixtures are propylene, dimethyl ether, propylene, propane, isobutane, ethane, and butane. They all have very low GWP [36] but belongs to A3 of safety group that means a higher flammability but a lower toxicity [35].

Table 2. Basic properties of the 61 zeotropic mixtures.

Identifying Number	Components	Composition/Mass %	Safety Group [35]	Type
R401A	R22/R152a/R124	53/13/34	A1	wet
R401B	R22/R152a/R124	61/11/28	A1	wet
R401C	R22/R152a/R124	33/15/52	A1	wet
R402A	R125/propane/R22	60/2/38	A1	wet
R402B	R125/propane/R22	38/2/60	A1	wet
R403A	propane/R22/R218	5/75/20	A2	wet
R430B	propane/R22/R218	5/56/39	A1	wet
R404A	R125/R134a/R143a	44/4/52	A1	wet
R405A	R22/R152a/R142b/RC318	45/7/5.5/42.5	A1	wet
R406A	R22/isobutane/R142b	55/4/41	A2	wet
R407A	R32/R125/R134a	20/40/40	A1	wet
R407B	R32/R125/R134a	10/70/20	A1	wet
R407C	R32/R125/R134a	23/25/52	A1	wet
R407D	R32/R125/R134a	15/15/70	A1	wet
R407E	R32/R125/R134a	25/15/60	A1	wet
R407F	R32/R125/R134a	30/30/40	A1	wet
R408A	R125/R143a/R22	7/46/47	A1	wet
R409A	R22/R124/R142b	60/25/15	A1	wet
R409B	R22/R124/R142b	65/25/10	A1	wet
R410A	R32/R125	50/50	A1	wet
R410B	R32/R125	45/55	A1	wet
R411A	propylene/R22/R152a	1.5/87.5/11	A2	wet
R411B	propylene/R22/R152a	3/94/3	A2	wet
R412A	R22/R218/R142b	70/5/25	A2	wet
R413A	R218/R134a/isobutane	9/88/3	A2	wet
R414A	R22/R124/isobutane/R142b	51/28.5/4/16.5	A1	wet
R414B	R22/R124/isobutane/R142b	50/39/1.5/9.5	A1	wet
R415A	R22/R152a	82/18	A2	wet
R415B	R22/R152a	25/75	A2	wet
R416A	R134a/R124/butane	59/39.5/1.5	A1	wet
R417A	R125/R134a/butane	46.6/50/3.4	A1	wet
R418A	propane/R22/R152a	1.5/96/2.5	A2	wet
R419A	R125/R134a/dimethylether	77/19/4	A2	wet
R420A	R134a/R142b	88/12	A1	wet
R421A	R125/R134a	58/42	A1	wet
R421B	R125/R134a	85/15	A1	wet
R422A	R125/R134a/isobutane	85.1/11.5/3.4	A1	wet
R422B	R125/R134a/isobutane	55/42/3	A1	wet
R422C	R125/R134a/isobutane	82/15/3	A1	wet
R422D	R125/R134a/isobutane	65.1/31.5/3.4	A1	wet
R423A	R134a/R227ea	52.5/47.5	A1	isentropic
R424A	R125/R134a/isobutane/butane/ isopentane	50.5/47/0.9/1/0.6	A1	wet
R425A	R32/R134a/R227ea	18.5/69.5/12	A1	wet
R426A	R125/R134a/butane/isopentane	5.1/93/1.3/0.6	A1	wet
R427A	R32/R125/R143a/R134a	15/25/10/50	A1	wet
R428A	R125/R143a/propane/isobutane	77.5/20/0.6/1.9	A1	wet
R429A	dimethylether/R152a/isobutane	60/10/30	A3	wet
R430A	R152a/isobutane	76/24	A3	wet
R431A	propane/R152a	71/29	A3	wet
R432A	propylene/dimethylether	80/20	A3	wet
R433A	propylene/propane	30/70	A3	wet
R434A	R125/R143a/R134a/isobutane	63.2/18/16/2.8	A1	wet
R435A	dimethylether/R152a	80/20	A3	wet
R436A	propane/isobutane	56/44	A3	isentropic
R436B	propane/isobutane	52/48	A3	isentropic
R437A	R125/R134a/butane/pentane	19.5/78.5/1.4/0.6	A1	wet
R438A	R32/R125/R134a/butane/ isopentane	8.5/45/44.2/1.7/0.6	A1	wet
R441A	ethane/propane/isobutane/butane	3.1/54.8/6/36.1	A3	wet
R442A	R32/R125/R134a/R152a/R227ea	31/31/30/3/5	A1	wet
R443A	propylene/propane/isobutane	55/40/5	A3	wet
R444A	R32/R152a/R1234ze(E)	12/5/83	A2L [1]	wet

[1] A2L is lower flammability refrigerants with a maximum burning velocity of ≤ 10 cm/s.

Table 3. Main thermodynamic and safety properties of six zeotropic mixtures.

Identifying Number	Critical Temperature (T_c)/K	The Extreme Temperature of Subcritical Region ($0.9T_C$)/K	Components	Composition /Mass %	Safety Group	Type
R432A	370.41	333.37	propylene/ dimethylether	80/20	A3	wet
R433A	367.57	330.81	propylene/ propane	30/70	A3	wet
R436A	389.04	350.14	propane/ isobutane	56/44	A3	isentropic
R436B	390.58	351.52	propane/ isobutane	52/48	A3	isentropic
R441A	391.62	352.46	ethane/propane/ isobutane/butane	3.1/54.8/6/36.1	A3	wet
R443A	369.16	332.24	propylene/propane /isobutane	55/40/5	A3	wet

3. Subcritical Cycle Analysis without Considering Isentropic Efficiency of Expander

First of all, it should be mentioned that the maximum operating temperature of single screw expander should not exceed 130 °C (400 K) due to the restriction of sealing material, lubricating oil, and starwheel material. The critical temperatures of all the zeotropic mixtures listed in Table 3 are lower than 400 K. Therefore, they are suitable candidates for zeotropic selection for ORC using single screw expander.

3.1. Thermodynamic Setting and Description

Generally speaking, an ORC usually runs in a subcritical rather than a transcritical or supercritical state due to the considerations of chemical stability and thermal stability of organic working fluid. Considering that four zeotropic mixtures are wet ones, $0.9T_c$ (critical temperature) is used to be the extreme temperature of subcritical region. That is to say, $0.9T_c$ (critical temperature) is also taken as the inlet temperature of the expander. On this basis, a subcritical ORC using single screw expander with wet zeotropic mixture as working fluid can be established and depicted by Figure 4.

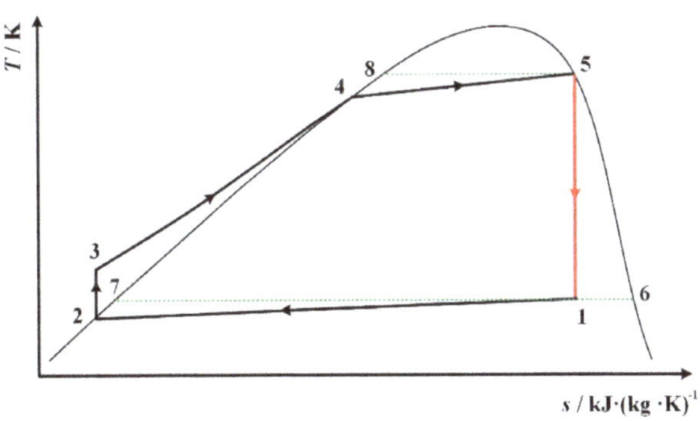

Figure 4. Subcritical ORC using single screw expander with wet zeotropic mixture as working fluid.

In a *T-s* diagram, a significant difference between dry (or isentropic) fluid and wet fluid is the existence of a point on which the entropy value reaches the maximum on saturated vapor curve ranging from normal boiling point to critical point. This point is located near the critical point and defined as the turning point. Its more detailed discussion can be found in paper [37]; other details about the role of this point can be seen elsewhere [38]. The turning point temperature is the limit of subcritical ORC which adopts turbo-type expander [37]. As for the two isentropic zeotropic mixtures, R436A and R436B, which are listed in Table 3, their turning point temperatures are 334.5 K and 342.5 K, respectively.

Because single screw expander that can tolerate vapor-liquid two-phase expansion is adopted in this study, $0.9T_c$ (critical temperature), which is a litter higher than turning point temperature, is still taken as the inlet temperature of the expander when wet zeotropic mixture is used as working fluid.

According to [18], 320 K and 290 K are the recommended condensation temperatures for the working fluids with high and low critical temperatures, respectively. These two condensation temperatures can be achieved by air cooling and water cooling. Accordingly, in order to make a detailed analysis, the thermodynamic performance of the above six zeotropic mixtures, including four wet ones and two isentropic ones, are calculated when the temperature of expander outlet varies from 290 K to 320 K.

The expander outlet states of four wet zeotropic mixtures are at vapor-liquid two-phase region. However, the expander outlet states of two isentropic zeotropic mixtures, R436A and R436B, are different. R436A's expander outlet is at vapor-liquid two-phase region, while R436B at superheated region. On this basis, a subcritical ORC using single screw expander with isentropic zeotropic mixture as working fluid can be established and depicted by Figure 5.

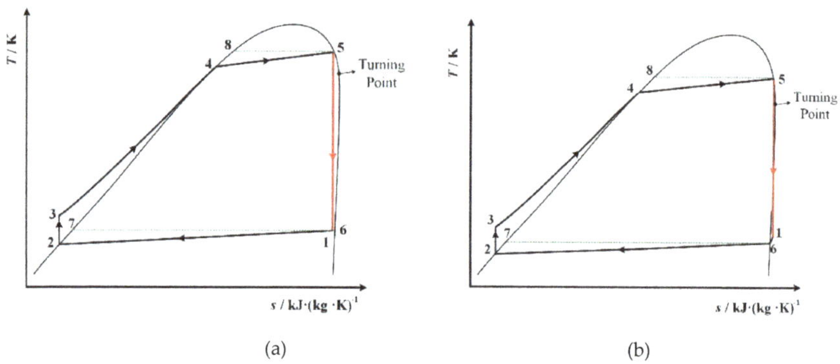

Figure 5. Subcritical ORC using single screw expander with isentropic zeotropic mixture as working fluid: (**a**)expander outlet at vapor-liquid two-phase region for R436A; (**b**)expander outlet at superheated region. for R436B.

All the state points in Figures 4 and 5 are described in Table 4.

Table 4. Description and determination of state points in Figures 4 and 5.

State Point	Description	Determination
1	expander outlet	$s_1 = s_5$
2	condenser outlet, at saturated liquid state	$p_2 = p_1$
3	pump outlet	$s_3 = s_2, p_3 = p_4$
4	at saturated liquid state	$p_4 = p_5$
5	expander inlet	$s_5 = s_1, T_5 = 0.9T_c$
6	at saturated vapor state	$T_6 = T_1$ for Figures 1 and 5a, $p_6 = p_1$ for Figure 5b
7	at saturated liquid state	$T_7 = T_6$
8	at saturated liquid state	$T_8 = T_5$

In Figures 4 and 5, net work is calculated by:

$$w_{net} = (h_5 - h_1) - (h_3 - h_2) \tag{1}$$

thermal efficiency is calculated by:

$$\eta = \frac{w_{net}}{q_e} = \frac{(h_5 - h_1) - (h_3 - h_2)}{h_5 - h_3} \quad (2)$$

heat exchange load of condenser is calculated by:

$$q_c = h_1 - h_2 \quad (3)$$

temperature glide in evaporator is calculated by:

$$\Delta T_e = T_5 - T_4 \quad (4)$$

temperature glide in condenser is calculated by:

$$\Delta T_c = T_6 - T_2 \quad (5)$$

and vapor quality for four wet zeotropic mixtures and one isentropic mixture(R436A) is calculated by:

$$x = \frac{s_1 - s_7}{s_6 - s_7} \quad (6)$$

In the above equations, h is enthalpy, s is entropy, T is temperature, w is work, q is heat exchange, η is thermal efficiency, c stands for condenser and condensation, e stands for evaporator and evaporation, and the numbers are the state points in the figure. In the following equations, h, s, T, w, q, η, and the numbers have the same meanings.

3.2. Results and Discussion

Table 5 lists the net work, thermal efficiency, heat exchange load of condenser, temperature glide in evaporator, temperature glide in condenser, and vapor quality of four wet zeotropic mixtures and two isentropic zeotropic mixtures when expander outlet temperature is known.

From the data listed in Table 5, it can be seen that among the six zeotropic mixtures, R441A has the best performance in net work, thermal efficiency, temperature glide in evaporator, and temperature glide in condenser, while R433A has the best performance in heat exchange load of condenser.

The vapor quality of four wet zeotropic mixtures increases with the increase of expander outlet temperature. As for two isentropic zeotropic mixtures, the vapor quality of R436A increases at first and then decreases. R436B does not have a vapor quality because its expander outlet is in the superheated region. The vapor quality of these five zeotropic mixtures is very high. In other words, they are very close to saturated vapor state.

For all six zeotropic mixtures, their four indexes, including net work, thermal efficiency, heat exchange load of condenser, and temperature glide in condenser, decrease with the increase of expander outlet temperature. The variation trends of these four indexes with the increase of expander outlet temperature are depicted in Figure 6.

Table 5. Thermodynamic performance of six zeotropic mixtures when expander outlet temperature is known.

Zeotropic Mixture	Expander Outlet Temperature T_1/K	Net Work/kJ·kg^{-1}	Thermal Efficiency/%	Heat Exchange Load of Condenser/kJ·kg^{-1}	Temperature Glide in Evaporator/K	Temperature Glide in Condenser/K	Vapor Quality
R432A	290	43.84	11.28	344.67	0.899	1.43	0.9450
	295	37.94	10.10	337.84		1.37	0.9491
	300	32.24	8.89	330.58		1.3	0.9534
	305	26.82	7.67	322.86		1.24	0.9579
	310	21.54	6.41	314.75		1.18	0.9628
	315	16.54	5.13	306.15		1.12	0.9685
	320	11.69	3.79	297.08		1.06	0.9749
R433A	290	40.11	10.66	336.24	0.263	0.32	0.9633
	295	34.37	9.46	328.91		0.32	0.9661
	300	28.89	8.25	321.13		0.31	0.9689
	305	23.61	7.02	312.89		0.31	0.9720
	310	18.55	5.75	304.16		0.3	0.9755
	315	13.74	4.45	294.92		0.29	0.9794
	320	9.13	3.10	285.13		0.28	0.9844
R436A	290	60.66	14.25	365.12	5.236	7.29	0.9973
	295	54.40	13.17	358.79		7.17	0.9980
	300	48.40	12.09	352.04		7.05	0.9985
	305	42.59	10.99	344.87		6.92	0.9987
	310	36.99	9.88	337.27		6.78	0.9987
	315	31.60	8.76	329.23		6.62	0.9983
	320	26.42	7.61	320.72		6.46	0.9979
R436B	290	62.39	14.58	365.65	5.342	7.48	N/A
	295	56.15	13.51	359.37		7.37	
	300	50.17	12.45	352.72		7.28	
	305	44.34	11.37	345.61		7.14	
	310	38.70	10.27	338.02		6.96	
	315	33.35	9.18	330.01		6.8	
	320	28.17	8.05	321.58		6.64	
R441A	290	69.38	14.68	403.38	12.958	18.54	0.9880
	295	62.56	13.60	397.28		18.20	0.9895
	300	55.94	12.52	390.79		17.85	0.9907
	305	49.56	11.44	383.83		17.48	0.9916
	310	43.40	10.34	376.44		17.09	0.9922
	315	37.47	9.23	368.59		16.68	0.9928
	320	31.75	8.10	360.25		16.26	0.9930
R443A	290	42.35	11.02	341.82	1.904	2.33	0.9580
	295	36.50	9.83	334.78		2.3	0.9611
	300	30.86	8.62	327.30		2.26	0.9644
	305	25.47	7.39	319.35		2.21	0.9678
	310	20.29	6.13	310.94		2.17	0.9717
	315	15.35	4.84	302.01		2.11	0.9760
	320	10.59	3.49	292.55		2.05	0.9812

From Figure 6a, it can be seen that R441A has the highest net work while R433A has the lowest. When the expander outlet temperature varies from 290 K to 320 K, R433A has the highest reduction, which is from 40.11 kJ/kg to 9.13 kJ/kg, a 77.24% decrease in net work. R441A has the lowest reduction, which is from 69.38 kJ/kg to 31.75 kJ/kg, a 54.24% decrease. From Figure 6b, it can be seen that R441A has the highest thermal efficiency and R433A has the lowest. When expander outlet temperature varies from 290 K to 320 K, R433A has the highest reduction, which is from 10.66% to 3.10%, a 70.89% decrease in thermal efficiency. R436B has the lowest reduction, which is from 14.58% to 8.05%, a 44.74% decrease. From Figure 6c, it can be seen that R441A has the highest heat exchange load of condenser and R433A has the lowest.

When the expander outlet temperature varies from 290 K to 320 K, R433A has the highest reduction, which is from 336.24 kJ/kg to 285.13 kJ/kg, a 15.20% decrease in heat exchange load of condenser. R441A has the lowest reduction, which is from 403.38 kJ/kg to 360.25 kJ/kg, a 10.69% decrease. From Figure 6d, it can be seen that R441A has the highest temperature glide in condenser while R433A has the lowest. When expander outlet temperature varies from 290 K to 320 K, R432A has the highest reduction, which is from 1.43 K to 1.06 K, a 25.87% decrease in temperature glide in condenser. R436B has the lowest reduction, which is from 7.48 K to 6.64 K, a 11.23% decrease.

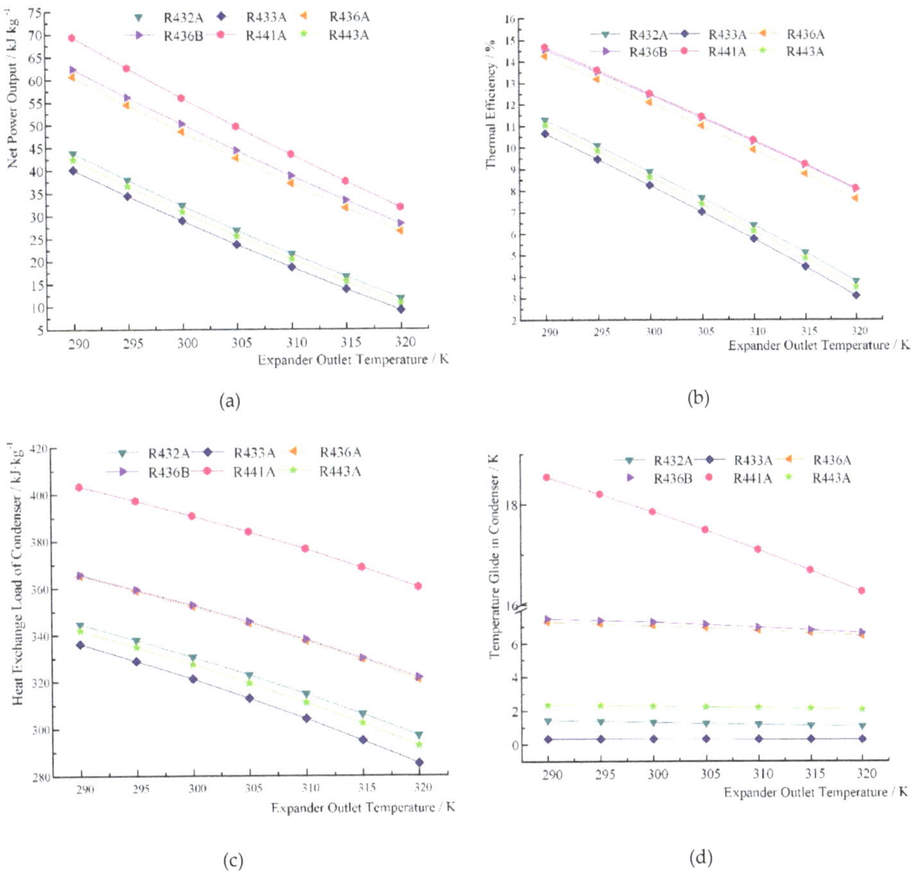

Figure 6. Variation trend of four indexes with the increase of expander outlet temperature: (**a**) net work; (**b**) thermal efficiency; (**c**) heat exchange load of condenser; (**d**) temperature glide in condenser.

From Table 5, it can be seen that the temperature glide in condenser of R441A ranges from 16 K to 18 K. If we assume a coolant with a temperature of 15 °C and a condenser pinch point temperature difference of 5 °C, all being realistic average values, then the minimum attainable working fluid expander exit temperature must be between 36 °C (15 °C + 5 °C + 16 °C = 36 °C =309 K) and 38 °C (15 °C + 5 °C + 18 °C = 38 °C = 311 K). Take the average value of these two temperatures, i.e., 310 K. Moreover, if we assume the expander exit temperature is 290 K, which is lower than the above average temperature, then in the case of R441a, would require the coolant temperature to be 290 K–17 K–5 K = 268 K(−5 °C), a temperature only attainable in arctic conditions. Therefore, 290 K is not a reasonable and feasible expander exit temperature for R441A. Based on the same principle, the other three exit temperatures, which are 295 K, 300 K, and 305 K, are not reasonable and feasible for R441A. Taken together, 310 K, 315 K, and 320 K are three reasonable and feasible expander exit temperatures for R441A.Because the expander inlet temperature is fixed at $0.9T_c$ for each zeotropic mixture, the temperature glide in evaporator keeps constant. The relation between the temperature glide in evaporator and the extreme temperature of subcritical region, which is $0.9T_c$ (critical temperature), is depicted in Figure 7. From Figure 7, it can be seen that R441A that has the highest critical temperature among all six zeotropic mixtures has the highest temperature glide in evaporator.

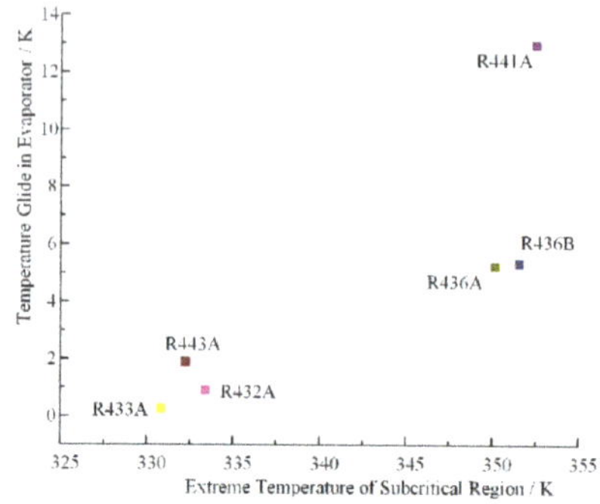

Figure 7. Relation between the temperature glide in condenser and the extreme temperature of subcritical region.

Table 6 lists the rank of six zeotropic mixtures. In Table 6, the heat exchange load of condenser is sorted from small to large, and the remaining items are sorted from large to small.

Table 6. The rank of 6 zeotropic mixtures.

Rank	Net Work	Thermal Efficiency	Heat Exchange Load of Condenser	Temperature Glide in Evaporator	Temperature Glide in Condenser
R441A	R441A	R433A	R441A	R441A	R441A
R436B	R436B	R443A	R436B	R436B	R436B
R436A	R436A	R432A	R436A	R436A	R436A
R432A	R432A	R436A	R443A	R443A	R432A
R443A	R443A	R436B	R432A	R432A	R443A
R433A	R433A	R441A	R433A	R433A	R433A

From the above analysis, it can be seen that R441A, which is a wet zeotropic mixture, can be selected as a suitable working fluid in subcritical ORC using single screw expander without considering isentropic efficiency of expander. It is suitable for both open and closed type heat source with a higher cost in heat exchanger. Its reasonable and feasible expander exit temperatures range from 310 K to 320 K.

4. Subcritical Cycle Analysis Considering Isentropic Efficiency of Expander

The above section has analyzed ideal subcritical ORC using single screw expander with zeotropic mixture as working fluid. It is based on isentropic efficiency of single screw expander is 100%. However, in practical application, its isentropic efficiency should be considered. Nowadays screw expanders show a much larger technical maturity than scroll and piston expanders [39]. The internal efficiency of single screw expander has exceeded 50% and the maximum is about 65% [25,40]. Therefore, 65% is used for analysis of single screw expander in this section.

4.1. Thermodynamic Setting and Description

Figure 8 depicts a subcritical ORC using a single screw expander with a wet zeotropic mixture as working fluid when considering the isentropic efficiency of the expander. Figure 9 depicts a subcritical ORC using a single screw expander with an isentropic zeotropic mixture as working fluid when considering the isentropic efficiency of the expander. In these figures, the blue dotted lines represent the expansion processes which has considered isentropic efficiency of expander. Here it should be noted that whether the isentropic efficiency is considered or not, the outlet temperature of the single screw expander remains the same. That is to say, $T_{1'} = T_1$ in Figures 8 and 9.

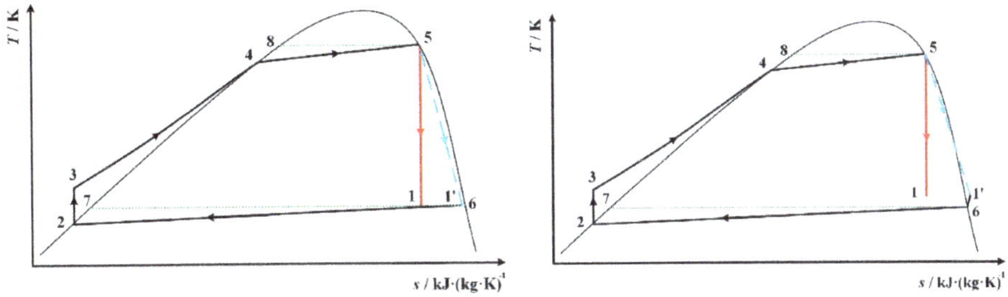

(a) (b)

Figure 8. Subcritical ORC using single screw expander with wet zeotropic mixture as working fluid when considering isentropic efficiency of expander: (**a**) expander outlet at vapor-liquid two-phase region; (**b**) expander outlet at superheated region.

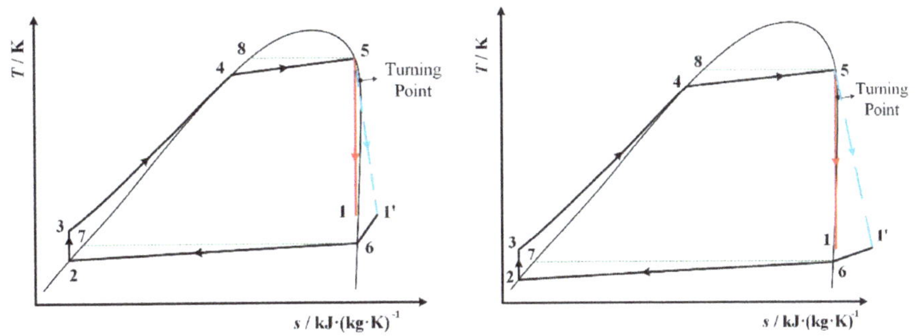

(a) (b)

Figure 9. Subcritical ORC using single screw expander with isentropic zeotropic mixture as working fluid when considering isentropic efficiency of expander: (**a**) expander outlet at superheated region for R436A; (**b**) expander outlet at superheated region for R436B.

All the state points in Figures 8 and 9 are described in Table 7.

Table 7. Description and determination of state points in Figures 8 and 9.

State Point	Description	Determination
1	expander outlet without considering isentropic efficiency	$s_1 = s_5$
1′	expander outlet considering isentropic efficiency	$T_{1'} = T_1, p_{1'} = p_6$
2	condenser outlet, at saturated liquid state	$p_2 = p_6$
3	pump outlet	$s_3 = s_2, p_3 = p_4$
4	at saturated liquid state	$p_4 = p_5$
5	expander inlet	$s_5 = s_1, T_5 = 0.9T_c$
6	At saturated vapor state	$T_6 = T_{1'}$ for Figure 5a, $p_6 = p_{1'}$ for Figures 8b and 9a and Figure 9b
7	at saturated liquid state	$T_7 = T_6$
8	at saturated liquid state	$T_8 = T_5$

In Figures 8 and 9, isentropic efficiency is calculated by:

$$\eta_{ex} = \frac{h_5 - h_1}{h_5 - h_1} \tag{7}$$

net work is calculated by:

$$w_{net} = (h_5 - h_{1'}) - (h_3 - h_2) \tag{8}$$

thermal efficiency is calculated by:

$$\eta = \frac{w_{net}}{q_e} = \frac{(h_5 - h_{1'}) - (h_3 - h_2)}{h_5 - h_3} \tag{9}$$

heat exchange load of condenser is calculated by:

$$q_c = h_{1'} - h_2 \tag{10}$$

temperature glide in evaporator is calculated by:

$$\Delta T_e = T_5 - T_4 \tag{11}$$

temperature glide in condenser is calculated by:

$$\Delta T_c = T_6 - T_2 \tag{12}$$

and vapor quality for the wet zeotropic mixture depicted in Figure 8a is calculated by:

$$x = \frac{s_{1'} - s_7}{s_6 - s_7} \tag{13}$$

4.2. Results and Discussion

Table 8 lists the net work, thermal efficiency, heat exchange load of condenser, temperature glide in the evaporator, temperature glide in the condenser, and vapor quality of four wet zeotropic mixtures and two isentropic zeotropic mixtures when the expander outlet temperature is known and considering isentropic efficiency of expander.

From the data listed in Table 8, it can be seen that when considering the isentropic efficiency of s single screw expander, the expansion process for different zeotropic mixtures is different. For R433A, its expander outlet is in the superheated region, shown by Figure 8b, when the outlet temperature varies from 290 K to 300 K and in the vapor-liquid two-phase region, shown by Figure 8a, when the outlet temperature varies from 305 K to 320 K. Similarly, for R443A, its expander outlet is in the superheated region, shown by Figure 8b, when the outlet temperature varies from 290 K to 295 K and in the vapor-liquid two-phase region, shown by Figure 8a, when the outlet temperature varies from 300 K to 320 K.

For two isentropic zeotropic mixtures, R436A and R436B, their expander outlets are always the superheated region, shown by Figure 9a,b, respectively.

For R441A that is a wet zeotropic mixture, its expander outlet is in the superheated region which is shown by Figure 8b. For another wet zeotropic mixture, R432A, its expander outlet is always in the vapor-liquid two-phase region, which is shown by Figure 8a.

As mentioned above, for R436A, R436B, and R441A, their expander outlets are always in the superheated region. Therefore, these three zeotropic mixtures need to be condensed from a superheated state to a saturated vapor state and then to a saturated liquid state. There is a very high temperature glide in the condenser. This makes the condenser outlet temperature extremely low. For example, when the expander outlet temperature is 310 K, R436A's condenser outlet temperature is 280.81 K and R436B's is 276.96 K. When the expander outlet temperature is 315 K, the condenser outlet temperature of R441A is only 278.77 K. These three condenser outlet temperatures are only slightly above 0 °C (273.15 K). It is difficult and uneconomical to reach this condensation temperature by using an air cooling or water cooling system, therefore, in subcritical ORC systems for R436A and R436B, the expander outlet temperature should be above 310 K. It should be above 315 K for the R441A system. This consideration confirms the rationality of the condensation temperature proposed in [18].

Table 8. Thermodynamic performance of six zeotropic mixtures when expander outlet temperature is known and considering isentropic efficiency of expander.

Zeotropic Mixture	Expander Outlet Temperature $T_{1'}$/K	Net Work/kJ·kg^{-1}	Thermal Efficiency/%	Heat Exchange Load of Condenser/kJ·kg^{-1}	Temperature Glide in Evaporator/K	Temperature Glide in Condenser/K	Vapor Quality
R432A	290	27.56	7.09	361.21	0.899	1.53	0.9903
	295	23.77	6.32	352.21		1.45	0.9895
	300	20.13	5.55	342.87		1.37	0.9888
	305	16.68	4.77	333.12		1.29	0.9882
	310	13.38	3.98	323.04		1.22	0.9881
	315	10.20	3.16	312.55		1.15	0.9886
	320	7.18	2.32	301.65		1.08	0.9899
R433A	290	25.07	6.55	357.54	0.263	0.34	N/A
	295	21.46	5.85	345.36		0.33	
	300	18.00	5.12	333.46		0.32	
	305	14.65	4.35	321.85		0.31	0.9998
	310	11.49	3.56	311.25		0.30	0.9983
	315	8.45	2.74	300.21		0.29	0.9970
	320	5.61	1.91	288.68		0.29	0.9967
R436A	290	N/A	N/A	N/A	5.236	N/A	N/A
	295						
	300						
	305						
	310						
	315	19.31	4.81	382.36		7.07	
	320	16.09	4.27	360.47		6.81	
R436B	290	N/A	N/A	N/A	5.342	N/A	N/A
	295						
	300						
	305						
	310						
	315	20.37	4.97	389.67		7.31	
	320	17.17	4.48	366.15		7.03	
R441A	290	N/A	N/A	N/A	12.958	N/A	N/A
	295						
	300						
	305						
	310						
	315						
	320	19.52	4.57	407.94		17.42	
R443A	290	26.55	6.84	361.71	1.904	2.57	N/A
	295	22.85	6.12	350.27		2.50	
	300	19.28	5.38	339.33		2.42	0.9996
	305	15.86	4.59	329.33		2.34	0.99784
	310	12.58	3.79	318.93		2.27	0.9965
	315	9.45	2.98	308.10		2.19	0.9956
	320	6.51	2.15	296.79		2.11	0.9953

Among all five evaluation indicators, temperature glide in the evaporator remains unchanged whether the isentropic efficiency of the expander is considered or not. The other four evaluation indicators, including net work, thermal efficiency, heat exchange load of condenser, and temperature glide in the condenser, decrease with the increase of expander outlet temperature when considering the isentropic efficiency of the expander. The variation trends of these four indicators with the increase of expander outlet temperature are depicted in Figure 10.

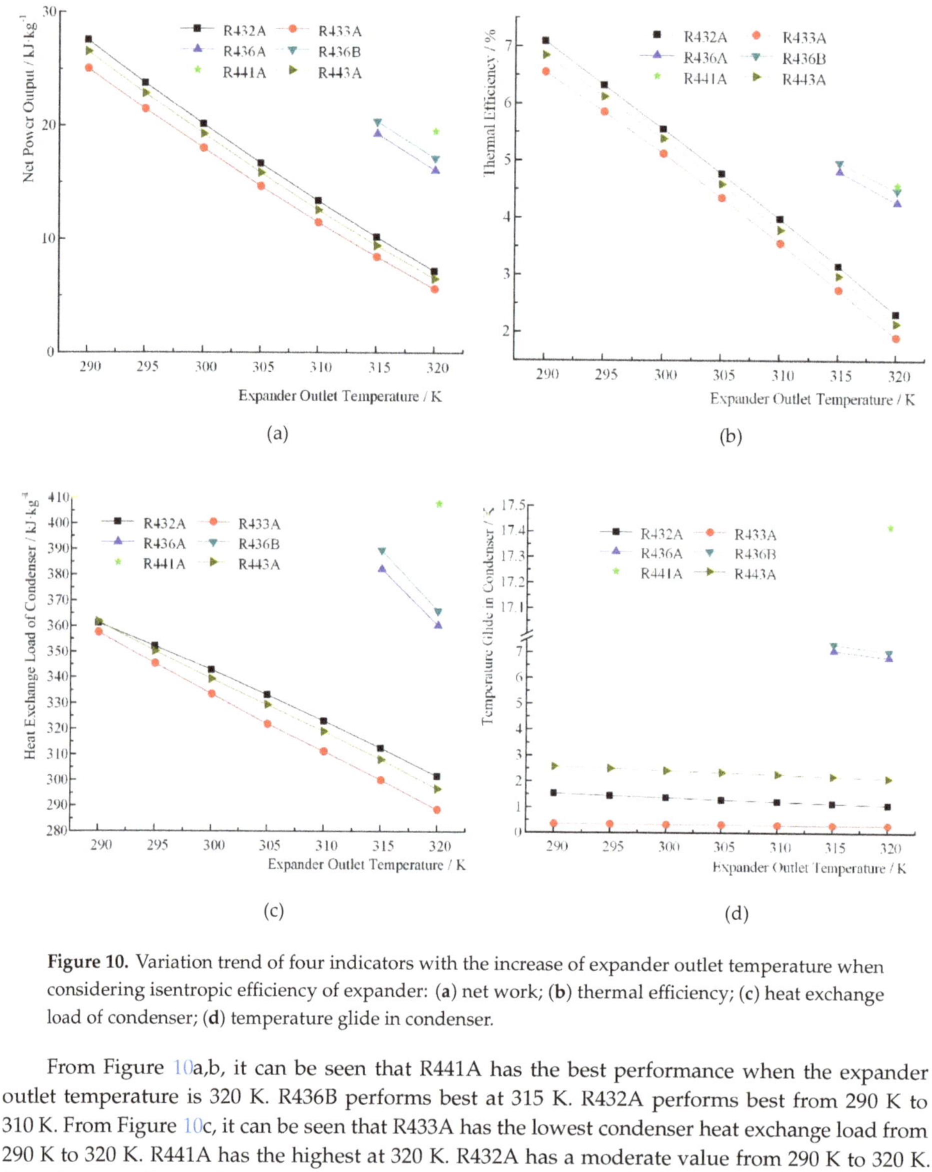

Figure 10. Variation trend of four indicators with the increase of expander outlet temperature when considering isentropic efficiency of expander: (**a**) net work; (**b**) thermal efficiency; (**c**) heat exchange load of condenser; (**d**) temperature glide in condenser.

From Figure 10a,b, it can be seen that R441A has the best performance when the expander outlet temperature is 320 K. R436B performs best at 315 K. R432A performs best from 290 K to 310 K. From Figure 10c, it can be seen that R433A has the lowest condenser heat exchange load from 290 K to 320 K. R441A has the highest at 320 K. R432A has a moderate value from 290 K to 320 K. From Figure 10d, it can be seen that R441A has the highest temperature glide in the condenser at 320 K. R432A has the moderate from 290 K to 320 K.

If we assume a coolant with a temperature of 15 °C and a condenser pinch point temperature difference of 5 °C, all being realistic average values, then the minimum attainable working fluid expander exit temperature must be 20 °C (15 °C + 5 °C, 293 K) plus the temperature glide in the condenser. Therefore, 290 K is not a reasonable and feasible expander outlet temperature for each working fluid.

The relation between vapor quality at expander outlet and expander outlet temperature is depicted by Figure 11. Without considering the isentropic efficiency of the expander, the vapor quality at the expander outlet increases with the increase of expander outlet temperature. When considering the isentropic efficiency of the expander, for R433A and R443A, their vapor quality at the expander outlet decreases with the increase of the expander outlet temperature. This is because the state of the expander outlet gradually changes from a superheated state to a two-phase state, whereas, R432A's expander outlet state is always in the two-phase region. There is a point with the longest distance from the outlet of the single screw expander to the saturated vapor curve. It is shown as "Point A" in Figure 12. With the increase of the expander outlet temperature, the vapor quality at the expander outlet decreases at first and then increases. This is because the distance from the outlet of the expander to the saturated vapor curve increases first and then decreases with the increase of the expander outlet temperature. Point A has the longest distance.

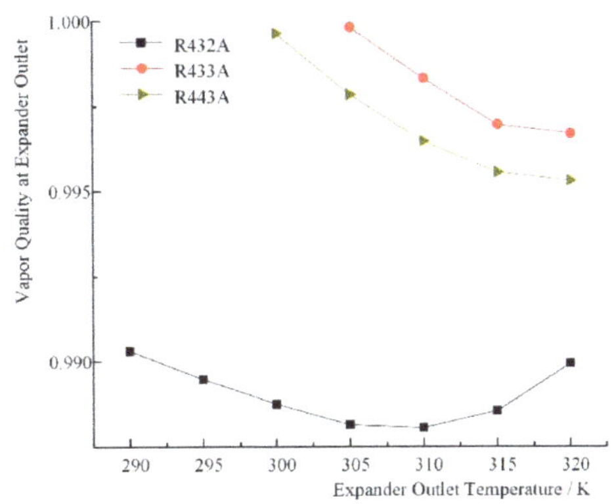

Figure 11. Relation between vapor quality at the expander outlet and the expander outlet temperature.

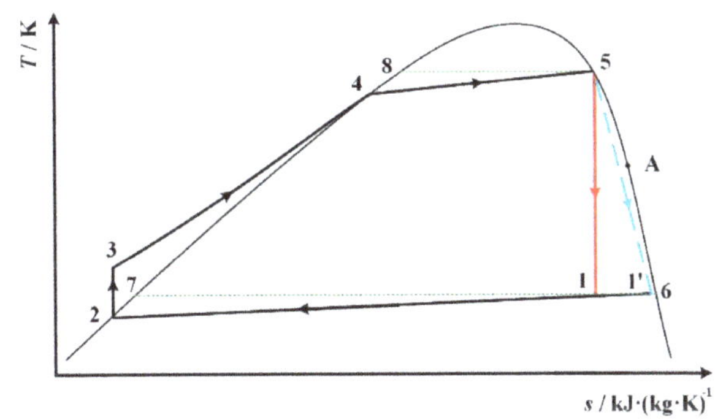

Figure 12. Point A with the longest distance between the outlet of the single screw expander and the saturated vapor curve.

Based on the above discussion and analysis, it can be seen that when considering the isentropic efficiency of a single screw expander, R441A with an expander outlet temperature of 320 K may be the suitable zeotropic mixture used for both open and close type heat sources. R436B may be selected for an expander outlet temperature of 315 K. R432A may be selected for an expander outlet temperature from 295 K to 310 K.

5. Conclusions

The organic Rankine cycle (ORC) is a popular and promising technology that has been widely studied and adopted in renewable and sustainable energy utilization and low-grade waste heat recovery. The use of zeotropic mixtures in ORC reduces the irreversibility in the evaporator and the condenser, thereby improving the thermodynamic performance. The selection of working fluid and expander are strongly interconnected. In order to make better use of the wet zeotropic mixture, a single screw expander, which is a novel expander with a tolerance of vapor-liquid two-phase expansion, was used in this paper for efficient utilization of the wet zeotropic mixtures listed in REFPROP 9.1 in a low-temperature subcritical ORC system.

Five indicators, namely net work, thermal efficiency, heat exchange load of condenser, temperature glide in evaporator, and temperature glide in the condenser, were used to analyze the performance of an ORC system with wet and isentropic zeotropic mixtures as working fluids.

Through calculation and analysis, it can be seen that R441A, which is a wet zeotropic mixture, can be selected as a suitable working fluid in subcritical ORC using a single screw expander without considering the isentropic efficiency of the expander. It is suitable for both open and closed type heat sources. Its reasonable and feasible expander exit temperatures range from 310 K to 320 K. When considering the isentropic efficiency of a single screw expander, R441A with an expander outlet temperature of 320 K may be the suitable zeotropic mixture for both open and close type heat sources. R436B may be selected for an expander outlet temperature of 315 K. R432A may be selected for an expander outlet temperature from 295 K to 310 K.

All three selected zeotropic mixtures belong to the A3 safety group classification which means a lower toxicity but a higher flammability. These three mixtures have very low GWP due to their components. In future practical applications, it is necessary to consider the ORC system complexity that may be caused by the use of the mixture.

Author Contributions: Conceptualization, X.Z.; Data curation, Y.Z., X.Z., and Z.L.; Formal analysis, X.Z. and Y.Z.; Funding acquisition, X.Z.; Methodology, X.Z. and Y.Z.; Resources, Y.W., C.M., and J.W.; Writing—original draft, X.Z.; Writing—review & editing, X.Z. and J.W. All authors have read and agreed to the published version of the manuscript.

Funding: This research was funded by the National Natural Science Foundation of China (Grant No.51506001) and Beijing Municipal Education Commission (KM201710005029). The authors gratefully acknowledge them for financial support of this work.

Conflicts of Interest: The authors declare no conflict of interest.

References

1. Imre, A.; Kustán, R.; Groniewsky, A. Thermodynamic selection of the optimal working fluid for organic Rankine cycles. *Energies* **2019**, *12*, 2028. [CrossRef]
2. Zhang, X.; He, M.; Wang, J. A new method used to evaluate organic working fluids. *Energy* **2014**, *67*, 363–369. [CrossRef]
3. Modi, A.; Haglind, F. A review of recent research on the use of zeotropic mixtures in power generation systems. *Energy Convers. Manag.* **2017**, *138*, 603–626. [CrossRef]
4. Radermacher, R. Advanced heat pump cycles using zeotropic refrigerant mixtures and solution circuits. *ASHRAE Trans.* **1986**, *92*, 2977.
5. Radermacher, R. Advanced Versions of Heat Pumps with Zeotropic Refrigerant Mixtures. *ASHRAE Trans.* **1991**, *92*, 52–59.
6. Weng, C. Experimental Study of Evaporative Heat Transfer for a Non-Azeotropic Refrigerant Blend at Low Temperature. Ph.D. Thesis, A Thesis Presented to the Faculty of the College of Engineering and Technology, Ohio University, Athens, OH, USA, 1990.
7. Abadi, G.B.; Kim, K.C. Investigation of organic Rankine cycles with zeotropic mixtures as a working fluid: Advantages and issues. *Renew. Sustain. Energy Rev.* **2017**, *73*, 1000–1013. [CrossRef]
8. Lecompte, S.; Ameel, B.; Ziviani, D.; Van den Broek, M.; De Paepe, M. Exergy analysis of zeotropic mixtures as working fluids in Organic Rankine Cycles. *Energy Convers. Manag.* **2014**, *85*, 727–739. [CrossRef]
9. Zhao, L.; Bao, J. Thermodynamic analysis of organic Rankine cycle using zeotropic mixtures. *Appl. Energy* **2014**, *130*, 748–756. [CrossRef]
10. Habka, M.; Ajib, S. Evaluation of mixtures performances in Organic Rankine Cycle when utilizing the geothermal water with and without cogeneration. *Appl. Energy* **2015**, *154*, 567–576. [CrossRef]

11. Deethayat, T.; Asanakham, A.; Kiatsiriroat, T. Performance analysis of low temperature organic Rankine cycle with zeotropic refrigerant by Figure of Merit (FOM). *Energy* **2016**, *96*, 96–102. [CrossRef]
12. Miao, Z.; Zhang, K.; Wang, M.; Xu, J. Thermodynamic selection criteria of zeotropic mixtures for subcritical organic Rankine cycle. *Energy* **2019**, *167*, 484–497. [CrossRef]
13. Battista, D.D.; Cipollone, R.; Villante, C.; Fornari, C.; Mauriello, M. The potential of mixtures of pure fluids in ORC-based power units fed by exhaust gases in Internal Combustion Engines. *Energy Procedia* **2016**, *101*, 1264–1271. [CrossRef]
14. Cipollone, R.; Bianchi, G.; Bartolomeo, M.D.; Battista, D.D.; Fatigati, F. Low grade thermal recovery based on trilateral flash cycles using recent pure fluids and mixtures. *Energy Procedia* **2017**, *123*, 289–296. [CrossRef]
15. Liu, B.; Chien, K.; Wang, C. Effect of working fluids on organic Rankine cycle for waste heat recovery. *Energy* **2004**, *29*, 1207–1217. [CrossRef]
16. Mago, P.J.; Chamra, L.M.; Srinivasan, K. An examination of regenerative organic Rankine cycles using dry fluids. *Appl. Therm. Eng.* **2008**, *28*, 998–1007. [CrossRef]
17. Hung, T.C. Waste heat recovery of organic Rankine cycle using dry fluids. *Energy Convers. Manag.* **2001**, *42*, 539–553. [CrossRef]
18. Chen, H.; Goswami, D.Y.; Stefanakos, E.K. A review of thermodynamic cycles and working fluids for the conversion of low-grade heat. *Renew. Sustain. Energy Rev.* **2010**, *14*, 3059–3067. [CrossRef]
19. Vélez, F.; Segovia, J.J.; Martín, M.C. A technical, economical and market review of organic Rankine cycles for the conversion of low-grade heat for power generation. *Renew. Sustain. Energy Rev.* **2012**, *16*, 4175–4189. [CrossRef]
20. Lemmon, E.W.; Huber, M.L.; McLinden, M.O. *NIST Standard Reference Database 23: Reference Fluid Thermodynamic and Transport Properties-REFPROP, Version 9.1*; National Institute of Standard Technology: Boulder, CO, USA, 2017.
21. Zhang, X.; Cao, M.; Yang, X.; Guo, H.; Wang, J. Economic analysis of organic Rankine cycle using R123 and R245fa as working fluids and a demonstration project report. *Appl. Sci.* **2019**, *9*, 288. [CrossRef]
22. Bao, J.; Zhao, L. A review of working fluid and expander selections for organic Rankine cycle. *Renew. Sustain. Energy Rev.* **2013**, *24*, 325–342. [CrossRef]
23. Lei, B.; Wang, W.; Wu, Y.; Ma, C.; Wang, J.; Zhang, L.; Li, C.; Zhao, Y.; Zhi, R. Development and experimental study on a single screw expander integrated into an Organic Rankine Cycle. *Energy* **2016**, *116*, 43–52. [CrossRef]
24. Wang, W.; Wu, Y.; Ma, C.; Liu, L.; Yu, J. Preliminary experimental study of single screw expander prototype. *Appl. Therm. Eng.* **2011**, *31*, 3684–3688. [CrossRef]
25. Lu, Y.; He, W.; Wu, Y.; Ji, W.; Ma, C.; Guo, H. Performance study on compressed air refrigeration system based on single screw expander. *Energy* **2013**, *55*, 762–768. [CrossRef]
26. Li, G.; Lei, B.; Wu, Y.; Zhi, R.; Zhao, Y.; Guo, Z.; Ma, C. Influence of inlet pressure and rotational speed on the performance of high pressure single screw expander prototype. *Energy* **2018**, *147*, 279–285. [CrossRef]
27. Wang, W.; Wu, Y.; Ma, C.; Xia, G.; Wang, J. Experimental study on the performance of single screw expanders by gap adjustment. *Energy* **2013**, *62*, 379–384. [CrossRef]
28. He, W.; Wu, Y.; Peng, Y.; Zhang, Y.; Ma, C.; Ma, G. Influence of intake pressure on the performance of single screw expander working with compressed air. *Appl. Therm. Eng.* **2013**, *51*, 662–669. [CrossRef]
29. Wang, W.; Wu, Y.; Xia, G.; Ma, C.; Ji, W.; Zhang, Y. Experimental study on the performance of the single screw expander prototype by optimizing configuration. In Proceedings of the ASME 2012 6th International Conference on Energy Sustainability collocated with the ASME 2012 10th International Conference on Fuel Cell Science, Engineering and Technology, San Diego, CA, USA, 23–26 July 2012; American Society of Mechanical Engineers Digital Collection: New York, NY, USA, 2012; pp. 1281–1286.
30. Xia, G.; Zhang, Y.; Wu, Y.; Ma, C.; Ji, W.; Liu, S.; Guo, H. Experimental study on the performance of single-screw expander with different inlet vapor dryness. *Appl. Therm. Eng.* **2015**, *87*, 34–40. [CrossRef]
31. Zhang, Y.; Wu, Y.; Xia, G.; Ma, C.; Ji, W.; Liu, S.; Yang, F. Development and experimental study on organic Rankine cycle system with single-screw expander for waste heat recovery from exhaust of diesel engine. *Energy* **2014**, *77*, 499–508. [CrossRef]
32. Zhang, X.; Zhang, Y.; Cao, M.; Wang, J.; Wu, Y.; Ma, C. Working Fluid Selection for Organic Rankine Cycle Using Single-Screw Expander. *Energies* **2019**, *12*, 3197. [CrossRef]

33. Yan, J.L. Thermodynamic principles and formulas for choosing working fluids and parameters in designing power plant of low temperature heat. *J. Eng. Thermophys.* **1982**, *3*, 1–7. (In Chinese)
34. He, C.; Liu, C.; Zhou, M. A new selection principle of working fluids for subcritical organic Rankine cycle coupling with different heat sources. *Energy* **2014**, *68*, 283–291. [CrossRef]
35. *ANSI/ASHRAE Standard 34-2016, Designation and Safety Classification of Refrigerants*; ASHRAE: Atlanta, GA, USA, 2016.
36. Intergovernmental Panel on Climate Change. In *Climate Change 2007—The Physical Science Basis*; Chapter 2: Changes in Atmospheric Constituents and in Radiative Forcing; Cambridge University Press: Cambridge, UK, 2007.
37. Zhang, X.; Zhang, C.; He, M.; Wang, J. Selection and Evaluation of Dry and Isentropic Organic Working Fluids Used in Organic Rankine Cycle Based on the Turning Point on Their Saturated Vapor Curves. *J. Therm. Sci.* **2019**, *28*, 643–658. [CrossRef]
38. Györke, G.; Deiters, U.K.; Groniewsky, A.; Lassu, I.; Imre, A.R. Novel Classification of Pure Working Fluids for Organic Rankine Cycle. *Energy* **2018**, *145*, 288–300. [CrossRef]
39. Lemort, V.; Guillaume, L.; Legros, A.; Declayea, S.; Quoilin, S. A comparison of piston, screw and scroll expanders for small scale Rankine cycle systems. In Proceedings of the 3rd International Conference on Microgeneration and Related Technologies, Naples, Italy, 15–17 April 2013.
40. Ziviani, D.; Gusev, S.; Lecompte, S.; Groll, E.A.; Braun, J.E.; Horton, W.T.; Broek, M.; De Paepe, M. Characterizing the performance of a single-screw expander in a small-scale organic Rankine cycle for waste heat recovery. *Appl. Energy* **2016**, *181*, 155–170. [CrossRef]

energies

MDPI

Article

Mapping of the Temperature–Entropy Diagrams of van der Waals Fluids

Attila R. Imre [1,2,*], Réka Kustán [2] and Axel Groniewsky [2]

1 Department of Energy Engineering, Faculty of Mechanical Engineering, Budapest University of Technology and Economics, Műegyetem rkp. 3, H-1111 Budapest, Hungary
2 Department of Thermohydraulics, Centre for Energy Research, POB. 49, H-1525 Budapest, Hungary; kustan@energia.bme.hu (R.K.); groniewsky@energia.bme.hu (A.G.)
* Correspondence: imreattila@energia.bme.hu

Received: 14 February 2020; Accepted: 16 March 2020; Published: 23 March 2020

Abstract: The shape of the temperature vs. specific entropy diagram of a working fluid is very important to understanding the behavior of fluid during the expansion phase of the organic Rankine cycle or similar processes. Traditional wet-dry-isentropic classifications of these materials are not sufficient; several materials remain unclassified or misclassified, while materials listed in the same class might show crucial differences. A novel classification, based on the characteristic points of the *T–s* diagrams was introduced recently, listing eight different classes. In this paper, we present a map of these classes for a model material, namely, the van der Waals fluid in reduced temperature (i.e., reduced molecular degree of freedom) space; the latter quantity is related to the molar isochoric specific heat. Although van der Waals fluid cannot be used to predict material properties quantitatively, the model gives a very good and proper qualitative description. Using this map, some peculiarities related to *T–s* diagrams of working fluids can be understood.

Keywords: adiabatic expansion; *T–s* diagram; working fluid classification; molecular degree of freedom

1. Introduction

Organic Rankine cycles (ORCs), where the traditional water/steam pair is replaced by an organic liquid/vapor pair, are very important in the utilization of low-temperature heat sources. In this way, even at relatively low temperatures, one can create organic vapor with sufficiently high pressure to drive turbines or expanders [1].

The selection of a proper working fluid is an important, multi-dimensional optimization problem [2]. Thermodynamical (e.g., efficiency), chemical (e.g., corrosion), biological (e.g., toxicity), environmental (e.g., global warming potential (GWP) and ozone depletion potential (ODP)), and other issues have to be considered with different weight [3,4]. For thermodynamic considerations, working fluids can be divided into several classes, depending on their behavior during the adiabatic expansion step of the ORC. Traditional classification uses three categories, namely wet, dry, and isentropic [2]. For wet fluids, starting the expansion from a saturated vapor state, the final state of an ideal (reversible adiabatic, i.e., isentropic) expansion is always a mixed, wet fluid state (droplets dispersed in the vapor). The presence of droplets should be avoided because they can damage the expanders; this can be done by the application of a superheater or droplet separator. For dry fluids, a similar expansion—except for those starting in the vicinity of the critical point—always ends in the dry, superheated vapor region. The presence of the superheated vapor requires greater cooling capacity from the condenser or the use of a recuperative or regenerative heat exchanger [5]. For isentropic fluids, the expansion would run along (or slightly above) the saturated vapor line, avoiding the previously mentioned problems. Unfortunately, isentropic fluids with an extended constant-entropy part on the saturated vapor branch do not exist; this part is always tilted or reverse S-shaped [6–8]. Concerning *T–s* (temperature–specific entropy) diagrams, the slope of the saturated vapor curve (the part of *T–s* diagram located on the high-entropy side of the critical point) is always negative for wet classes, always positive (except a tiny negative region close to the critical point) for dry ones, and theoretically would be infinite (except a tiny negative region close to the critical point) for isentropic ones [9–11].

There are several disadvantages to this traditional classification: two of them are shown in Figure 1. In Figure 1a,b, one can see the two schematic *T–s* (temperature–specific entropy) diagrams of two hypothetical dry fluids. One can clearly see that although according to the previously mentioned criterium (isentropic expansion from a saturated vapor state terminated in dry vapor region, shown by solid red arrows) both fluids are dry, still, they have considerable differences. The liquid and vapor parts of the saturation curve are separated by the critical point (C), having the liquid part on the low-entropy, and the vapor part on the high-entropy side. One of the important difference is that in the case of Figure 1a, there is a theoretical possibility to fully transform the fluid from liquid to vapor and vice versa by pure ideal compression or expansion; In contrast, in the case of Figure 1b, this transformation will be only partial (see dashed arrows). In the first case, the diagram is strongly tilted and therefore for low-temperature saturated vapor states, the entropies are lower than for high-temperature liquid phases (giving the possibility to the afore-mentioned isentropic vapor-to-liquid or liquid-to-vapor transitions), while in the second case, on the entropy-scale, vapor states are always above the liquid ones, disabling the system to fully vaporize or liquefy in an adiabatic step. This example shows that the dry class should be divided into at least two subclasses, depending on the relative position (on the entropy scale) of the critical point and the end-point of the saturated vapor branch.

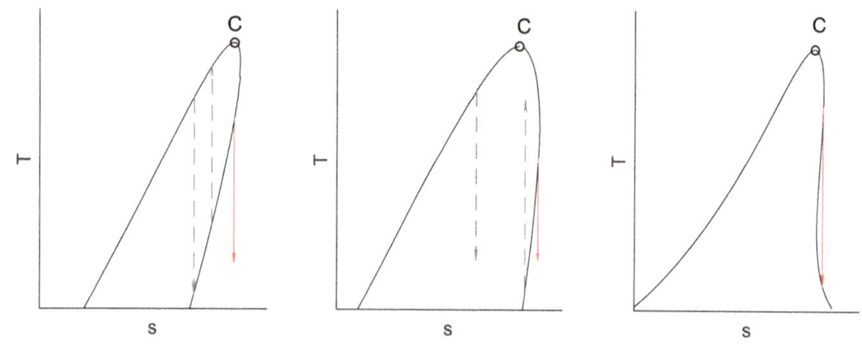

(a) **(b)** **(c)**

Figure 1. Schematic temperature–specific entropy (*T–s*) diagrams of two well-distinguishable dry working fluid subclasses (**a**,**b**) with a previously unclassifiable type (**c**). Some relevant expansion and compression routes are shown by arrows; the importance of these routes are described in the text.

The other problem is that for real working fluids, one can easily find a *T–s* diagram with a special shape, not accurately covered by the dry-wet-isentropic classes. This type is shown in Figure 1c. Traditionally, some of the fluids showing this shape were considered as dry (regarding only the upper, high-temperature part) or isentropic (only in cases where the inverse S-shape of the saturated vapor curve was so flat that it was considered as an almost straight, vertical line) while a lot of them were wrongly classified or not classified at all [12]. Practically, fluids showing this behavior can be forced into the isentropic class because it is possible to have an ideal adiabatic "saturated vapor to saturated vapor" expansion (see full arrow) [6]. However, while for theoretical isentropic working fluids (where part of the saturated vapor curve would be a straight, vertical line) it is possible to expand from any temperature to any other in a reversible adiabatic manner (at least within the temperature range, where "isentropicity" would be true), for these reverse S-shaped ones, it is possible to do so only between certain temperature pairs (connecting two points with a vertical line, like on Figure 1c). For a given fluid, these pairs—considered as starting/end-points of ideal expansion steps—can be represented by a curve on an upper vs. lower temperature diagram. These diagrams can be used to select working fluid for a given heat sink/heat source pair [6]. These fluids, due to the existence of isentropic "saturated vapor to saturated vapor" expansion steps (red arrow, Figure 1c), can be referred as "real isentropic" working fluids (distinguishing them from the idealized isentropic ones).

To solve for these shortcomings, a novel classification was introduced [12] based on the entropy sequence of characteristic points on the *T*–*s* diagrams. These points (shown in Figure 2) were the two end-points of the curve (marked as A for the low-entropy and Z as the high-entropy side), the critical point (C), and two local extrema on the saturated vapor part, a maximum (M) and a minimum (N). A, C, and Z points exist for all materials, while M and N exist only for the dry or the reverse S-shaped ones. Since that A, C, and Z points are present for all materials, they are called primary characteristic points, while M and N (being present only for the non-wet working fluids) are the secondary characteristic points.

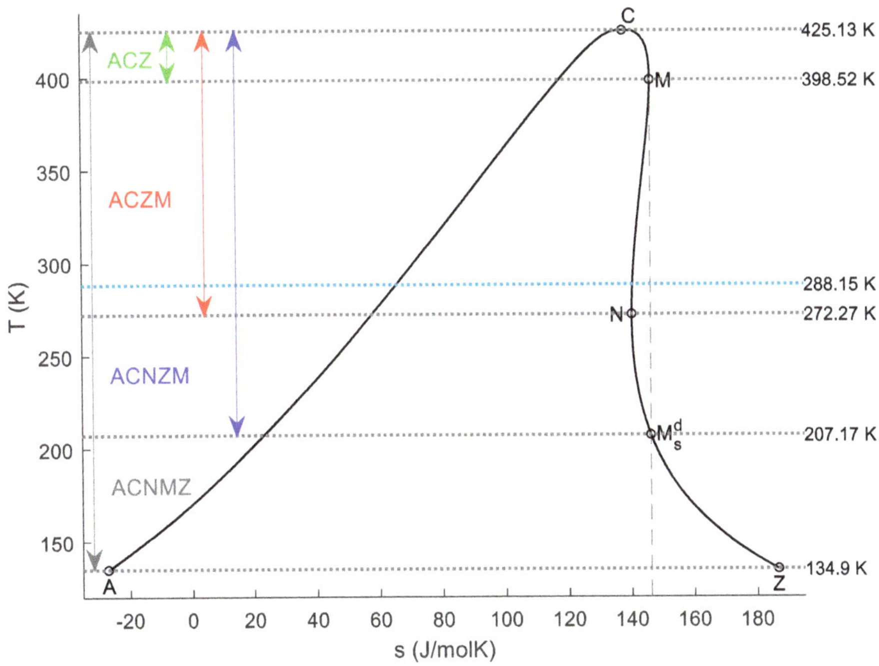

Figure 2. Temperature–specific entropy (*T*–*s*) diagram of butane, showing the classification in the full fluid-range (from the triple point to the critical point) as well as the classification (and related characteristic points) in various confined temperature ranges (colored arrows and letters). The blue dotted line shows ambient temperature (15 °C). *T*–*s* data taken from the NIST Webbook [13].

From these five characteristic points, one can theoretically construct 3! + 4! + 5! = 6 + 24 + 120 sequences containing only the primary points (A, C, and Z) or the primaries plus M (A, C, Z, and M) or finally all the five points (A, C, Z, M, and N); it is an enormous number to replace the previously used three classes. Fortunately, due to some constraints (for example, the entropy of point C is always bigger than that for point A, or the entropy of point M is always above the entropy values for points C and N) one can define only eight possible sequences, giving one subclass for wet (ACZ), two subclasses for dry (ACZM and AZCM; the first one is demonstrated in Figure 1b, the second one is in Figure 1a), and five subclasses for the "real isentropic" sequence (ANCMZ, ACNMZ, ANZCM,

ANCZM, and ACNZM) [12]. These subclasses are shown in Appendix A, while the classification of some real. pure working fluids (only for the ones having accurate T–s data in the NIST Chemistry Webbook [13]) can be found in the Supplementary Data Section of reference [12].

Taking the two end-points (A and Z) as the absolute end-points of the T–s curves, i.e., locating them to the triple point, the obtained class will be an absolute material property, just like critical temperature, molar mass, etc. On the other hand, these classifications sometimes are not very user-friendly, because the temperature of the triple point sometimes is much below the temperature range applicable in usual ORC processes. Therefore, we introduced a few more characteristic points, called ternary characteristic points [14]. They are created by projecting the primary and secondary points to the T and s axes; here, we are using only the latter ones. The projection line extended along the whole temperature scale can cross the original T–s diagram, defining the ternary points by these intersections. Due to the nature of the projection, the entropy or the temperature of these points coincide with the entropy or temperature of the corresponding primary or secondary characteristic points. For example, by projecting point C to the entropy axis, it might cross the original diagram once or twice, depending on the class of the fluid. These ternary points are marked as C_s^d and C_s^{dd}; s index marks the axis where the points are projected, d marks the position of the new ternary point compared to the original point; therefore these points (being d for down) are on temperatures below the temperature of point C. Finally, the number of the upper indices are marking the intersection nearer (one index) or farther (double index) from the original characteristic point. The importance of these points and the temperature-dependent part of this novel classification method is demonstrated through the case of butane (Figure 2); T–s data taken from the NIST Webbook [13].

Butane is a type ACNMZ working fluid; the fluid range extends from the triple-point temperature, 134.90 K (−138.25 °C) to the critical temperature 425.13 K (151.98 °C). One can easily realize that in ORCs it is quite unlikely that butane would expand down to −138 °C, or even in the vicinity, except for some cryogenic applications [15]. Therefore, it is important to know how the classification can change by fixing the lower end to an environment-given minimal temperature (like 288.15 K = 15 °C), instead of fixing it to the material-given triple-point temperature. In this way, one might realize, that while a working fluid (butane, in this case) might be an ACNMZ-type fluid, in a confined temperature range, it can emulate other types, even an ACZ one. These transitions can be seen in Figure 2. For this fluid class (ACNMZ), the entropy of point N is above the entropy of all saturated liquid states, therefore, projecting point C to the entropy axis would not yield any intersection with the saturated vapor curve. The projection line for N would yield one intersection (N_s^u) close to point C, but as will be seen later, this point has no relevance here, therefore it is not shown. The projection line of M (dashed) crosses the saturated vapor line in a point, which is referred to here as M_s^d. There are also some dotted lines, marking the temperatures of the primary and secondary points, as well as for this new ternary point. One further temperature is also marked (288.15 K = 15 °C) as the ambient temperature.

As was already mentioned, considering the full fluid range between the triple-point temperature (134.90 K) and critical temperature (425.13 K), butane is a type ACNMZ working fluid; this is shown in Figure 2 with grey characters and the corresponding temperature range is marked with a grey arrow. When, for some practical reason, we are interested only in higher temperature ranges, the situation changes. Butane remains ACNMZ type only up to the temperature of point M_s^d, 207.17 K. This means that when we are interested in expansion properties for butane only above this temperature, the butane behaves like an ACNZM-type fluid (instead of the original ACNMZ class), considering the temperature range between points M_s^d and C; this range is shown by a double-headed blue-colored arrow. This happens because the entropy of the new end-point of the saturated vapor curve is now lower than the entropy of point M, therefore, Z exchanges places with M in the sequence (MZ ending turns to ZM, while the first three letters remain intact). In some cases [12], these temperature-dependent end-points are marked with stars (Z*; and because A is also connected to the same temperature, A*); therefore the new class would be A*CNZ*M to show that it is a temperature-dependent class: For the sake of simplicity, we are omitting the stars for now. Increasing the temperature of the new end-point further, reaching point N (at 272.27 K) would cause the butane to behave like a dry, ACZM-type fluid because above that temperature, point N (local entropy minimum) falls below the new end-point (and letter N disappears from the sequence). Increasing the temperature even further, the next class-change happens by reaching point M (398.52 K). Hence, when the operation temperature is between point N to point C (272.27 K to 425.13 K) butane behaves as an ACZM-type fluid (temperature range marked by a red arrow), from point M to point C it behaves like a wet, ACZ-type one (range is shown by a green arrow), although these temperatures (above point M) are hardly used in ORC applications [16]. An extra temperature (288.15 K = 15 °C) marks a plausible ambient temperature, which can be taken as the lower cycle temperature, for example, for an air-cooled geothermal ORC unit. In that case, butane will have a dry, ACZM-type behavior.

Here, one can see a change of sequences: ACNMZ→ACNZM→ACZM→ACZ. For other real materials, these sequences can be different; for example, for water and carbon dioxide, which are ACZ types, the classification remains unchanged by increasing the lower end temperature, while for dodecafluoropentane, the sequence is AZCM→ACZM→ACZ (*T*–*s* data for substances are taken from the NIST Webbook [13]). Using data for model fluids (for example, simple van der Waals fluid [5] or Redlich–Kwong fluid [17]) or simply using geometrically correct schematic representations [12], one can make the temperature-dependent sequence change for all the eight types. The figures explaining the changes are in the Appendix A, together with the list of temperature-dependent sequences. For better visibility of the saturated vapor curves, the original point A (the end-point of the liquid saturation curve, taken at the triple-point temperature) is not always shown in the Appendix A, being a very low entropy value. It is interesting that while an ANCMZ-type fluid can emulate an AZCM- or ACZM-type, an ACNMZ-type can only imitate an ACZM-type and cannot be forced to behave as an AZCM-type. Additionally, all types turn to ACZ-type when the lower end-point is sufficiently close to the critical point. In this way, it is possible to make a tree-like graph, marking the various potential routes of the class changes; this tree can be seen in Figure 3, color codes are identical with the colors used in Figure 2.

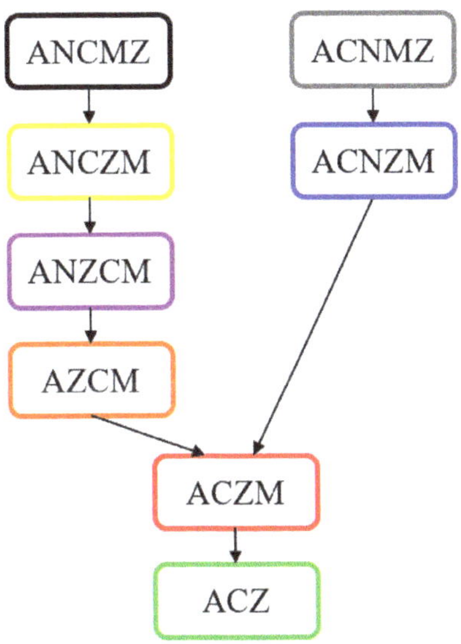

Figure 3. The potential changes in the classes by changing the lower end-point temperature. Color codes are identical with the colors of Figures 2 and 4.

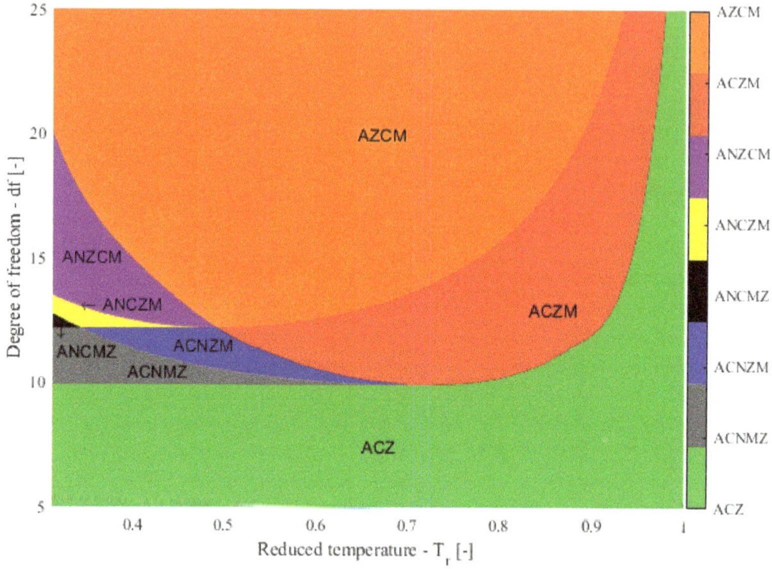

Figure 4. Reduced temperature vs. molecular degree of freedom for simple, one-component van der Waals fluids. Three-letter, four-letter, and five-letter parts show various wet, dry, and real isentropic subclasses.

Figure 3 show a double-branched tree: ACZ–ACZM is the trunk and AZCM–ANZCM–ANCZM–ANCMZ and ACNZM–ACNMZ are two branches. This means that from any given working fluid, only a maximum of five other classes can be emulated by changing the lower end-point temperature, i.e., the longest chain can be a maximum of six steps long.

The number six also turns up in another problem. Checking all pure fluids included in the NIST Webbook [13], real fluids fit into only six classes (see examples in Table 1), i.e., fluids showing types ACZM and ACNZM in their whole fluid range do not exist (or at least not in this set of 72 fluids). Also, representing simple fluids with van der Waals equations of state [5] and changing the molecular complexity, reflected in the molecular degree of freedom and the molar isochoric heat capacity, one can see a smooth transition from ACZ-type through ACNMZ–ANCMZ–ANCZM–ANZCM to AZCM-type; an animated GIF about the transition can be seen on Wikipedia [18]. This six-class long sequence differs from the previous one; but while in the previous case one fixed diagram (i.e., one material with a given molecular complexity) was studied in various temperatures, in the second case, the situation was reversed, the temperature (as end-point temperature) was kept as a constant and the molecular degree of freedom was changed from low to high. In the case of simulated fluids, a lower end-point temperature has to be appointed by us since the van der Waals and Redlich–Kwong EoS are unable to predict phase transitions with solid phases, i.e., the lower (triple point) temperature has to be an externally given quantity. For the van der Waal case [5] it was given as $0.31 \times T_c$. In the case of real fluids, the triple-point temperatures are not located in the same reduced temperature, but there is a very soft rule of thumb [19,20] that places the triple point for most materials between $0.3 \times T_c$ and $0.4\ T_c$, i.e., these values are also roughly the same (although as it can be seen for real working fluids, the triple-point temperatures for some halogenated alkanes are closer to $0.5 \times T_c$ [5]).

Table 1. Some real examples for various working fluid classes, for more examples, see reference [12].

Class	Materials
ACZ	water, methane, ethane
ACNMZ	butane, freon 115
ANCMZ	pentane, toluene
ANCZM	benzene
ANZCM	cyclohexane, freon 113
AZCM	freon 114, dodecane, octafluorocyclobutane

These examples show that some classes are connected (i.e., it is possible to go from one to the other, either by slightly changing the temperature or the molecular complexity), while others are distinct. To understand this phenomenon, as well as to help us to design novel working fluids, a mapping of different classes in the reduced temperature vs. molecular complexity space is presented here. We use simple van der Waals fluid for mapping; although the van der Waals equation of states is not able to describe material properties quantitatively, qualitatively it can describe most of the existing phenomenon. A good example to show the abilities of the van der Waals equation is the case of global phase diagrams, where different types of binary van der Waals mixtures were mapped, describing the general phase properties of almost all non-aqueous binary mixtures [21–23].

2. Results: Mapping of Working Fluids

For the classification, *T*–*s* diagrams of van der Waals fluids with different molecular degrees of freedom marked as *df* (from 3 to 30, with steps 0.01) were calculated in the reduced temperature scale (from 0.31 to 1) with 0.0003 steps. Details of the calculation can be found in references [5,17]. In that simple model, chain molecules were considered, where the maximal molecular degree of freedom is 3**n*, and *n* is the number of atoms in the molecule, but to obtain smoother transition between classes, this variable was taken as a continuous one. In the given model, the degree of freedom was assumed to be temperature-independent; hence, molar isochoric heat capacity was taken as $c_{Vm} = \frac{df}{2}R$ [5]. Then, primary and secondary points were determined numerically using a self-made MATLAB code. Finally, using the entropy values of these points, classes were determined and plotted in a reduced temperature–degree of freedom diagram. The relevant part of the diagram (molecular degrees of freedom values between 5 and 25) is shown in Figure 4, while a magnified part, showing special points of the map, is shown in Figure 5.

All the eight classes can be found on the map (Figure 4), although some of them (like ACZ or AZCM) cover bigger areas, while others, like ANCMZ (small black triangle-shaped area) cover only a smaller portion. Some of the classes are neighboring ones, which means that by changing the temperature or molecular complexity by a small fraction, a fluid can step from one class to another (like ACZM and ACZ). Going back to butane (Figure 2), taking a lower end-point of 398.51 K, the system is emulating ACZM-type, while taking this temperature as 398.53 K, the type turns to ACZ. Other classes are more distinct, for example, from ACZM (red) it is impossible to reach ANCMZ (black) by a small step.

Borders can be divided into five, or rather four and a half parts. Two of these border lines are linear (straight line), at a fixed degree of freedom (*df* = 9.92 and *df* = 12.19). Two (and a half) others are curved ones; the first runs between green and red then blue and red, finally between violet and orange; this curve has another branch (taken as a half curve) going out from the minimum of the previous curve, separating the grey and blue, then the yellow and black regions. The first part of the border is already known and describes the location of N and M points; it has a minimum, existing even for real materials, which can be used as a rule of thumb to distinguish between wet and non-wet working fluids, based on their measured molar isochoric heat capacity in a given (fixed-value) reduced temperature [14,24]. Although there are several rules of thumb to predict at least the basic properties of working fluids, most of them are not accurate or very complex. Only a handful of them work with nearly 100% accuracy, predicting the wetness/dryness of the fluid [24–26] by using simple quantities for these correlations. The last line is also a curved one, running between the red-orange and the yellow-violet regions. Straight borders terminate when reaching any of the curved ones; these common points are the minima of the curved ones. Correlations (polynomial fits) of these borders can be found in Appendix B.

There are three special points on the map, where four or five classes meet; they are numbered in Figure 5, which is a magnification of the relevant part of Figure 4, showing these multiple points.

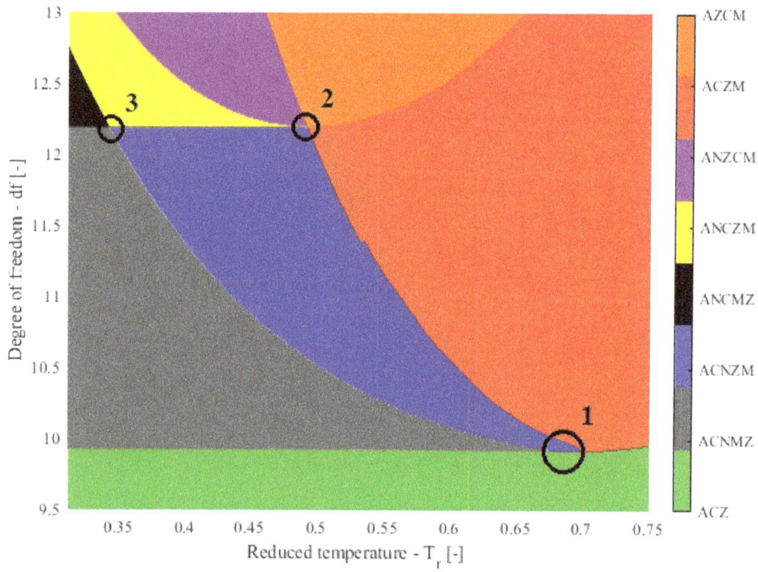

Figure 5. Magnified part of Figure 4 showing the multiple points of the map.

In point 1, ACZM, ACNMZ, ACNZM, and ACZ classes form a quadruple point (although ACNMZ, the grey one, seems to be terminated earlier, this is only due to numerical noise). In the immediate vicinity of this point (marked by 1 in Figure 5) one can find four different classes. It is possible to move from ACZM (red) to ACNMZ (grey) by changing the molecular complexity, characterized here by the molecular degree of freedom (*df*) while keeping the reduced temperature constant in a two-step or in a direct, one-step process. During a two-step process, from the ACZM (red) class, one has to reach the ACNZM class (blue) first; this happens when *T*(N)—the temperature of point N—which was originally hidden below the lower temperature limit, related to the solid-liquid-vapor triple-point temperature, suddenly pops up to the accessible temperature range. During the second step, from this ACNZM (blue) class, by further changing the molecular complexity, one can reach the ACNMZ class (grey); this occurs when the entropies for M and Z (*s*(M) and *s*(Z)) switch places. This happens when the so-called ideal gas-part contribution [5] increases which increasing *df* and shifts point M to higher entropy values. Based on the analysis of this two-step process, one can easily describe what happens during the one-step process of ACZM (red) to ACNMZ (grey). Put simply, the two independent phenomena, namely the appearance of point M (when *T*(N) reaches *T*(Z)) and the switch of M and Z (when *s*(Z) goes above *s*(M)), happen in the same place in the df–T_r space. Similar coincidences are responsible for the other quadruple point (marked as 3 in Figure 5) as well as for the existence of the quintuple point where five classes can exist within the immediate vicinity of point 2 (Figure 5).

The locations of the multiple points shown in Figures 4 and 5 are shown in Table 2, giving their reduced temperature and molecular degree of freedom coordinates.

Table 2. The location of the multiple points shown in Figure 5.

Point	Location
1	$df = 9.92$ $T_r = 0.6991$–0.6739
2	$df = 12.19$ $T_r = 0.4903$–0.4918
3	$df = 12.19$, $T_r = 0.406$–0.3412

3. Discussion: Explanation of Various Phenomena with the Map

We investigated some of our previous findings on the map. First, as we already stated, all classes turn to ACZ when the new end-point is sufficiently close to the critical point. The critical point is represented here by $T_r = 1$, and it is clearly seen that close to this value, only the ACZ (green) class exists.

Second, we were able to find only six classes among real materials (when the lower end-point is fixed to the triple point, around $T_r = 0.3$–0.4) as well as for van der Waals fluids (where we fixed the end-point to $T_r = 0.3$). It can be seen, that at $T_r = 0.3$, there are only six classes, ACZ (green), ACNMZ (grey), ANCMZ (black), ANCZM (yellow), ANZCM (violet), and finally, AZCM (orange). Also, one can see that by fixing the temperature elsewhere (for example at $T_r = 0.4$), we would see a similar, but not the same sequence: ANCMZ (black) would be replaced by ACNZM (blue). Fixing the end-point temperature around $T_r = 0.55$, the sequence would be shorter, ACZ–ACNMZ–ACNZM–ACZM–AZCM, while at an even higher temperature ($T_r = 0.8$), it would be a three-step process, ACZ–ACZM–AZCM.

As was seen previously, the classes found among real materials were identical with the classes found for van der Waals fluids with the end-point fixed at $T_r = 0.3$, but since this value has nothing to do with real triple-point temperatures (except for when using the soft rule of thumb mentioned above), we believe that this exact equality is just a lucky coincidence. However, it can still show that no more than six classes can exist when the lower end-point is fixed in a given reduced temperature.

Third, by fixing the molecular degree of freedom (i.e., taking one, fixed material) and by changing the end-point temperature from the real, triple point related one up to the critical point, the maximal number of classes emulated during this process is six (see the longer route for Figure 3). In Figure 4, by choosing a degree of freedom value around 12.5 one can see an ANCMZ (black) – ANCZM (yellow) – ANZCM (violet) – AZCM (orange) – ACZM (red) – ACZ (green), six-step sequence, while even slightly below this value, at $df = 12$, one can see the four-step sequence, namely ACNMZ (grey) – ACNZM (blue) – ACZM (red) – ACZ (green), down to $df = 9.92$. Below this value, only ACZ-type fluids exist. With more complex molecules (above $df = 20$) one can see only dry or wet ones (AZCM-ACZM-ACZ) for $T_r > 0.3$. The two lines separating the map into these regions are the straight-line borders at $df = 9.92$ and $df = 12.19$.

Fourth, as can be seen in Figure 3, three of the eight classes seem to be archetypes: ACZ, ANCMZ, and ACNMZ, the three classes at the end of this graph. ACZ occupies the high-temperature end, i.e., approaching $T_r = 1$, where all fluids turn to ACZ-type. On the other end, going to $T_r = 0$, fluids with $df \leq 9.92$ (small molecules) are ACZ-type (green) and fluids with $9.92 \leq df \leq 12.19$ (medium-size molecules) are ACNMZ-type (grey). One can assume that the black region runs up (like the green region on the high-temperature side), and therefore at the low-temperature limit, all fluids with $df > 12.19$ (longer molecules) would be class ANCMZ. In this way, $T_r = 1$ and $T_r = 0$ limits would be occupied only by these three archetypes.

One might ask that if the calculations for van der Waals fluids were started at a more reduced temperature (like $T_r = 0.4$), would the black ANCMZ region disappear? Would it not be possible to find hidden ninth or even tenth working fluid classes in temperature or degree of freedom regions not shown in Figure 4? The answer is no; it has been shown in reference [12] that these are the only classes; the existence of other classes are theoretically impossible. On the other hand, one cannot rule out that any of these eight classes could re-appear in another T_r–df region (most probably on the high-df region, representing very complex molecules), i.e., the existence of two separate regions with the same type (i.e., two distinct blue regions) cannot be fully ruled out. This problem might require further study.

It can be assumed that using other equations of states, the map can differ to some extent; these differences can be qualitative or quantitative. According to some previous results [17], the classes are probably located in a similar order when the Redlich–Kwong equation is used, but the whole diagram is shifted to higher *df* values. Further studies to describe these differences are in progress.

4. Conclusions and Summary

For the organic Rankine cycle, the selection of working fluid is crucial; it can affect not only the output (efficiency, power) but also the layout of the equipment used for the realization of the cycle [5,27]; this is especially true for the selection of the expander [28]. For some fluids, one may use only a basic layout of pump-evaporator-expander-condenser, while for other fluids, the use of extra parts, like a droplet separator, superheater, or recuperative heat exchanger may be necessary. The type of the layout can be correlated with the shape of the temperature–specific entropy (*T*–*s*) diagram of the given working fluid, more specifically with the shape (and location) of the saturated vapor branch of this curve. Based on the shape, one can divide existing as well as potential working fluids into various classes. The traditional three-class (wet-dry-isentropic) classification is not sufficient, because several materials can remain unclassified or misclassified, while materials listed in the same class might show crucial differences. Therefore, recently, a more detailed, eight-class classification was proposed; this new classification is compatible with the traditional ones, but can solve the problems associated with traditional ones.

Classes can be handled as material properties, using the full fluid part of the *T*–*s* diagrams (from triple point to critical point). Sometimes the lower limit (triple point) is located in temperatures hardly used in ORC processes; therefore, one shall also define a lower end-point temperature classification when the end-point of the saturated liquid and vapor branches are taken in a temperature value more practical than triple-point temperature. In the new, limited temperature range (new end-point to critical point) a working fluid belonging to a given class can emulate the general properties of another class.

Using a model fluid (like van der Waals fluid), material property can also be a smoothly changeable property, where transitions from one to another class can be demonstrated; this can help us to design novel working fluids.

In this paper, we give the map of a van der Waals fluid in a reduced temperature vs. molecular degree of freedom diagram. All the eight novel classes can be located on the map; peculiarities (for example, the fact that from real fluids we were not able to find examples for two of these classes; or, the existence of two temperature-dependent classification routes) can be easily explained using the map. Three multiple points (where more than two classes meet in a given reduced temperature and molecular degree of freedom value) were located. Although the map based on the van der Waals equation of state cannot give quantitatively correct predictions, qualitatively it can describe all known phenomena.

Author Contributions: Conceptualization, A.R.I.; analysis: R.K. and A.G.; writing: A.R.I., A.G., and R.K. All authors have read and agreed to the published version of the manuscript.

Funding: This work was performed in the frame of the FIEK_16-1-2016-0007 project, implemented with the support provided from the National Research, Development and Innovation Fund of Hungary, financed under the FIEK_16 funding scheme. Some parts of the research reported in this paper have been supported by the National Research, Development and Innovation Fund (TUDFO/51757/2019-ITM), Thematic Excellence Program. R.K. has been supported by the ÚNKP-19-3 New National Excellence Program of the Ministry for Innovation and Technology.

Conflicts of Interest: The authors declare no conflict of interest.

Appendix A Temperature-Dependent Classification for Various Working Fluid Classes

As was already mentioned in the main text, when fixing the two end-points (A and Z) to the triple-point temperature as the ultimate lower-end of the liquid and vapor states, the classes are material properties (concerning their whole range for fluid phases, for example, propane is an ACZ-type wet working fluid, while dodecane is an AZCM-type dry working fluid). For practical purposes, sometimes it is more useful to have an application-related end-point for the *T*–*s* diagram; in ORC applications it can be the minimal temperature of the cycle, related to the temperature of the heat-sink, often very close to the environmental temperature, but in other cases, like in cryogenic cycles, it can differ from that significantly. Using a new end-point temperature in the new fluid range (stretching from the new end-point temperature to the critical temperature), the working fluid, originally listed in one class, can emulate the behavior of another class. The transition happens by reaching some of the primary, secondary, or ternary characteristic points. The transitions for the eight theoretical classes are shown in Figure A1.

(a)	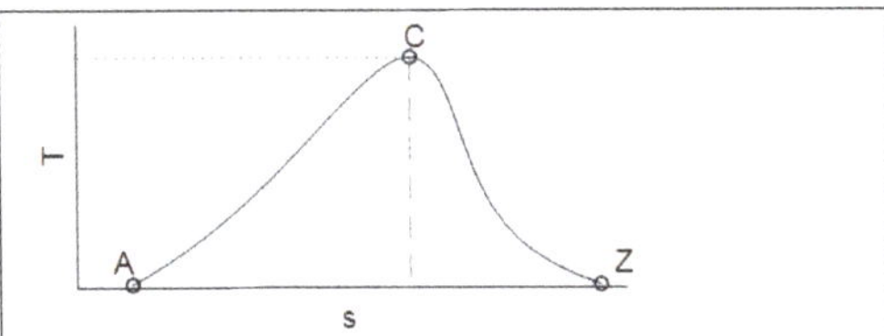	**ACZ**
(b)	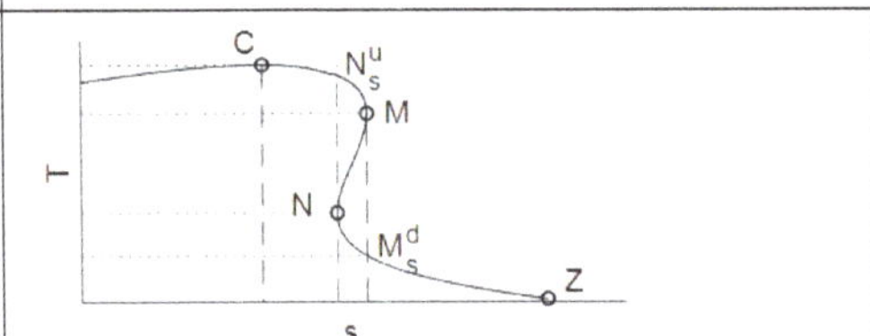	**ACNMZ** T(C) – T(M): ACZ T(M) – T(N): ACZM T(N) – T(M_s^d): ACNZM T(M_s^d) – T(Z): ACNMZ
(c)	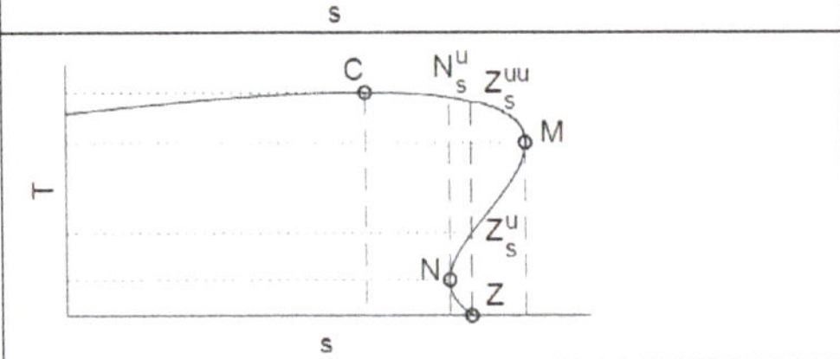	**ACNZM** T(C) – T(M): ACZ T(M) – T(N): ACZM T(N) – T(Z): ACNZM
(d)	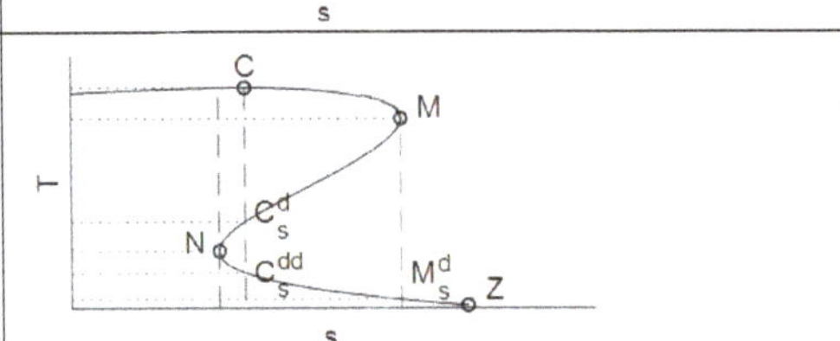	**ANCMZ** T(C) – T(M): ACZ T(M) – T(C_s^d): ACZM T(C_s^d) – T(N): AZCM T(N) – T(C_s^{dd}): ANZCM T(C_s^{dd}) – T(M_s^d): ANCZM T(M_s^d) – T(Z): ANCMZ
(e)	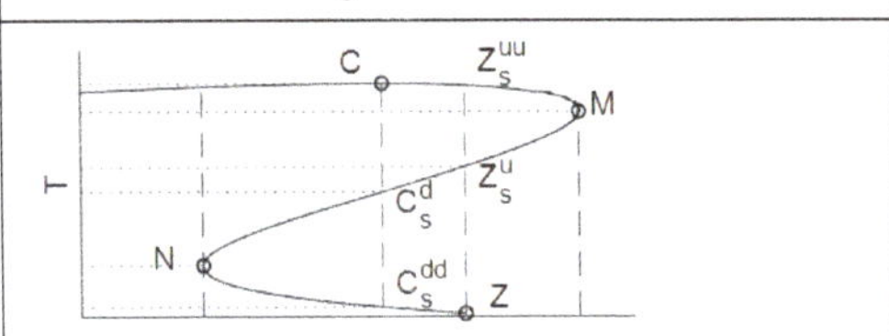	**ANCZM** T(C) – T(M): ACZ T(M) – T(C_s^d): ACZM T(C_s^d) – T(N): AZCM T(N) – T(C_s^{dd}): ANZCM T(C_s^{dd}) – T(Z): ANCZM

Figure A1. *Cont.*

(f)	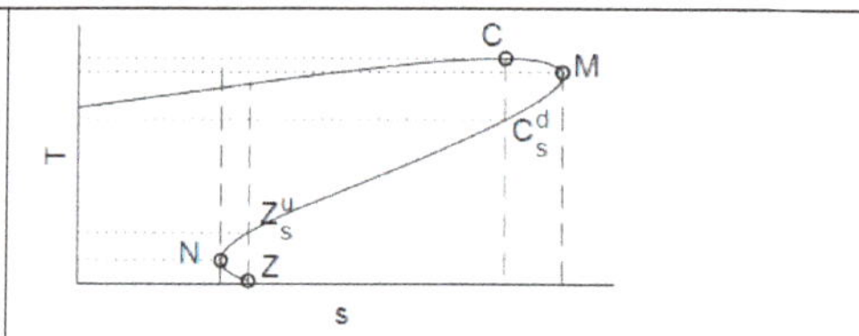	**ANZCM** T(C) – T(M): ACZ T(M) – T(C_s^d): ACZM T(C_s^d) – T(N): AZCM T(N) – T(Z): ANZCM
(g)	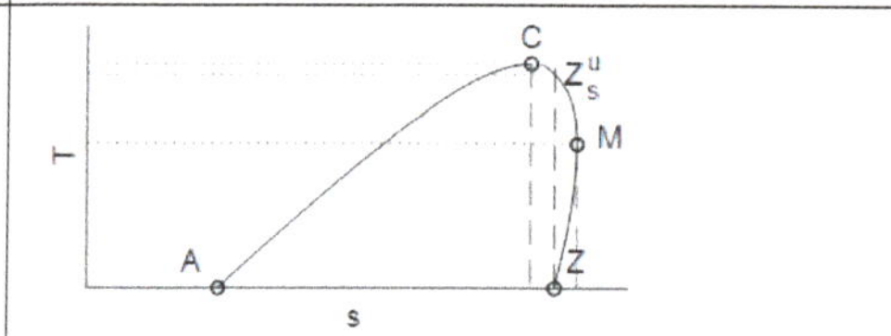	**ACZM** T(C) – T(M): ACZ T(M) – T(Z): ACZM
(h)	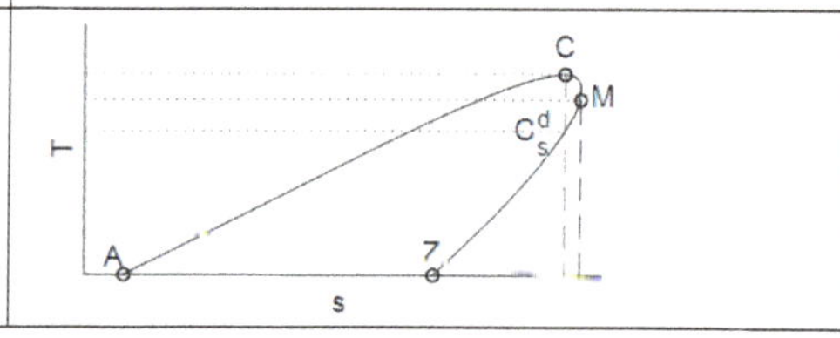	**AZCM** T(C) – T(M): ACZ T(M) – T(C_s^d): ACZM T(C_s^d) – T(Z): AZCM

Figure A1. Temperature-dependent classification scheme for various working fluid classes.

For better visibility of the saturated vapor curves, the original point A (the end-point of the liquid saturation curve, taken at the triple-point temperature) is not always shown, as it has very low entropy. Bold letters mark the original classification for the full fluid scale (from the triple point to critical point), while normal capital letters mark the location of the new end-point and the corresponding new class.

Appendix B Correlations of Border Lines Separating Various Classes

Borders between various classes can be described by polynomial fits. The fitting curves are shown and marked by Arabic numbers in Figure A2. Fitting parameters and other relevant data (like R-square and the range of fitting (given as minimal T_r, maximal T_r)) are also listed.

Curve I

Correlation function:

$f(x) = p1^* x^9 + p2^* x^8 + p3^* x^7 + p4^* x^6 + p5^* x^5 + p6^* x^4 + p7^* x^3 + p8^* x^2 + p9^* x + p10$

Coefficients (with 95% confidence bounds):

p1 = 4.357e + 05 (4.252e + 05, 4.463e + 05)

p2 = −2.278e + 06 (−2.34e + 06, −2.216e + 06)

p3 = 5.175e + 06 (5.016e + 06, 5.333e + 06)

p4 = −6.692e + 06 (−6.926e + 06, −6.459e + 06)

p5 = 5.416e + 06 (5.199e + 06, 5.632e + 06)

p6 = −2.835e + 06 (−2.966e + 06, −2.703e + 06)

p7 = 9.554e + 05 (9.035e + 05, 1.007e + 06)

p8 = −1.985e + 05 (−2.115e + 05, −1.856e + 05)

p9 = 2.276e + 04 (2.092e + 04, 2.46e + 04)

p10 = −1046 (−1160, −931.7)

Goodness of fit: SSE: 3.992; R-square: 1; Adjusted R-square: 1; RMSE: 0.03643

Temperature range: $T_r = [0.3, 1]$

Curve II

Correlation function:

$f(x) = p1^* x^4 + p2^* x^3 + p3^* x^2 + p4^* x + p5$

Coefficients (with 95% confidence bounds):
p1 = 70.49 (66.39, 74.6)
p2 = −180.2 (−188.2, −172.2)
p3 = 184.5 (178.8, 190.2)
p4 = −90.66 (−92.44, −88.88)
p5 = 27.84 (27.63, 28.04)
Goodness of fit: SSE: 0.002998; R-square: 1; Adjusted R-square: 1; RMSE: 0.003272
Temperature range: T_r = [0.3, 0.6739]

Curve III

Correlation function:
$f(x) = p1^* x^9 + p2^* x^8 + p3^* x^7 + p4^* x^6 + p5^* x^5 + p6^* x^4 + p7^* x^3 + p8^* x^2 + p9^* x + p10$
Coefficients (with 95% confidence bounds):
p1 = 3.649e + 05 (3.404e + 05, 3.895e + 05)
p2 = −1.981e + 06 (−2.122e + 06, −1.841e + 06)
p3 = 4.706e + 06 (4.353e + 06, 5.059e + 06)
p4 = −6.41e + 06 (−6.917e + 06, −5.903e + 06)
p5 = 5.512e + 06 (5.052e + 06, 5.971e + 06)
p6 = −3.099e + 06 (−3.372e + 06, −2.827e + 06)
p7 = 1.139e + 06 (1.033e + 06, 1.244e + 06)
p8 = −2.633e + 05 (−2.89e + 05, −2.376e + 05)
p9 = 3.471e + 04 (3.114e + 04, 3.829e + 04)
p10 = −1972 (−2188, −1756)
Goodness of fit: SSE: 5.841; R-square: 0.9999; Adjusted R-square: 0.9999; RMSE: 0.05536
Temperature range: T_r = [0.3, 1]

Line IV

Linear at df = 9.92
Temperature range: T_r = [0.3, 0.6991]

Line V

Linear at df = 12.19
Temperature range: T_r = [0.3, 0.4918]

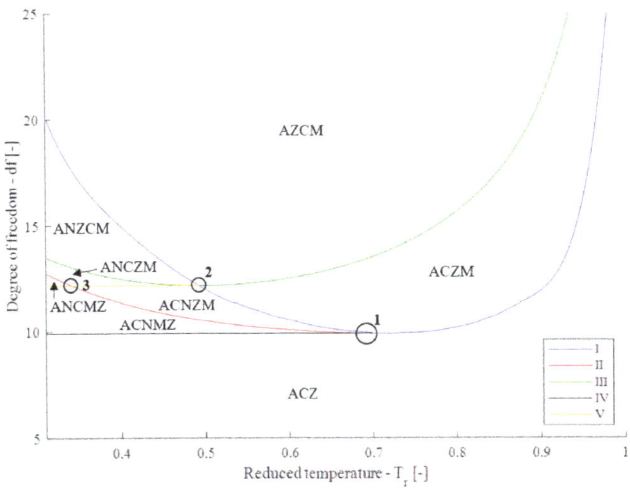

Figure A2. Correlation curves (df vs. T_r) for the different borders of the maps shown in Figure 4.

References

1. Macchi, E.; Astolfi, M. *Organic Rankine Cycle (ORC) Power Systems: Technologies and Applications*; Elsevier-Woodhead Publishing: Duxford, UK, 2016.
2. Qiu, G. Selection of working fluids for micro-CHP systems with ORC. *Renew. Energy* **2012**, *48*, 565–570. [CrossRef]
3. Chen, H.; Goswami, D.Y.; Stefanakos, E.K. A review of thermodynamic cycles and working fluids for the conversion of low-grade heat. *Renew. Sustain. Energy Rev.* **2010**, *14*, 3059–3067. [CrossRef]
4. Bao, J.; Zhao, L. A review of working fluid and expander selections for Organic Rankine Cycle. *Renew. Sustain. Energy Rev.* **2013**, *24*, 325–342. [CrossRef]
5. Groniewsky, A.; Györke, G.; Imre, A.R. Description of wet-to-dry transition in model ORC working fluids. *Appl. Therm. Eng.* **2017**, *125*, 963–971. [CrossRef]
6. Imre, A.R.; Kustán, R.; Groniewsky, A. Thermodynamic Selection of the Optimal Working Fluid for Organic Rankine Cycles. *Energies* **2019**, *12*, 2028. [CrossRef]
7. White, J.A.; Velasco, S. A Simple Semiempirical Method for Predicting the Temperature-Entropy Saturation Curve of Pure Fluids. *Ind. Eng. Chem. Res.* **2019**, *58*, 1038–1043. [CrossRef]
8. White, J.A.; Velasco, S. Approximating the Temperature—Entropy Saturation Curve of ORC Working Fluids From the Ideal Gas Isobaric Heat Capacity. *Energies* **2019**, *12*, 3266. [CrossRef]
9. Morrison, G. The shape of the temperature-entropy saturation boundary. *Int. J. Refrig.* **1994**, *17*, 494. [CrossRef]
10. Garrido, J.M.; Quinteros-Lama, H.; Mejía, A.; Wisniak, J.; Segura, H. A rigorous approach for predicting the slope and curvature of the temperature-entropy saturation boundary of pure fluids. *Energy* **2012**, *45*, 888–899. [CrossRef]
11. Rivera-Alvarez, A.; Abakporo, O.I.; Osorio, J.; Hovsapian, R.; Ordonez, J.C. Predicting the slope of the temperature-entropy vapor saturation curve for working fluid selection based on Lee-Kesler modeling. *Ind. Eng. Chem. Res.* **2020**, *59*, 956–969. [CrossRef]
12. Györke, G.; Deiters, U.K.; Groniewsky, A.; Lassu, I.; Imre, A.R. Novel Classification of Pure Working Fluids for Organic Rankine Cycle. *Energy* **2018**, *145*, 288–300. [CrossRef]
13. NIST Chemistry WebBook. NIST Standard Reference Database Number 69. 2018. Available online: http://webbook.nist.gov/chemistry/ (accessed on 1 November 2019).
14. Imre, A.R.; Groniewsky, A. Various ways of adiabatic expansion in Organic Rankine Cycle (ORC) and in Trilateral Flash Cycle (TFC). *Z. Phys. Chem.* **2019**, *233*, 577–594. [CrossRef]
15. Sadaghiani, M.S.; Ahmadi, M.H.; Mehrpooya, M.; Pourfayaz, P.; Feidt, M. Process development and thermodynamic analysis of a novel power generation plant driven by geothermal energy with liquefied natural gas as its heat sink. *Appl. Therm. Eng.* **2018**, *133*, 645–658. [CrossRef]
16. Zhang, X.; Zhang, C.; He, M.; Wang, J. Selection and Evaluation of Dry and Isentropic Organic Working Fluids Used in Organic Rankine Cycle Based on the Turning Point on Their Saturated Vapor Curves. *J. Therm. Sci.* **2019**, *28*, 643–658. [CrossRef]
17. Groniewsky, A.; Imre, A.R. Prediction of the ORC working fluid's temperature-entropy saturation boundary using Redlich-Kwong equation of state. *Entropy* **2018**, *20*, 93. [CrossRef]
18. Working Fluid Selection Page. *Wikipedia*. Available online: https://en.wikipedia.org/wiki/Working_fluid_selection (accessed on 1 November 2019).
19. Taft, R.; Stareck, J. Relationship between Melting-Points, Normal Boiling-Points and Critical Temperatures. *J. Phys. Chem.* **1930**, *34*, 2307–2317. [CrossRef]
20. Bowden, S.T. A Corrected Guldberg Rule. *Nature* **1954**, *174*, 613–614. [CrossRef]
21. Polishuk, I.; Wisniak, J.; Segura, H.; Yelash, L.V.; Kraska, T. Prediction of the critical locus in binary mixtures using equation of state—II. Investigation of van der Waals-type and Carnahan-Starling-type equations of state. *Fluid Phase Equilib.* **2000**, *172*, 1–26. [CrossRef]
22. Van Konynenburg, P.H.; Scott, R.L. Critical lines and phase equilibriums in binary Van der Waals mixtures. *Philos. Trans. R. Soc. Lond. Ser. A Math. Phys. Eng. Sci.* **1980**, *298*, 495–540.
23. Deiters, U.K.; Kraska, T. *High-Pressure Fluid Phase Equilibria—Phenomenology and Computation*; Elsevier: Amsterdam, the Netherlands, 2012.

24. Györke, G.; Groniewsky, A.; Imre, A.R. A simple method to find new dry and isentropic working fluids for Organic Rankine Cycle. *Energies* **2019**, *12*, 480. [CrossRef]
25. Albornoz, J.; Mejía, A.; Quinteros-Lama, H.; Garrido, J.M. A rigorous and accurate approach for predicting the wet-to-dry transition for working mixtures in organic Rankine cycles. *Energy* **2018**, *156*, 509–519. [CrossRef]
26. White, J.A.; Velasco, S. Characterizing wet and dry fluids in temperature-entropy diagrams. *Energy* **2018**, *154*, 269–276. [CrossRef]
27. Xia, X.X.; Qi, W.Z.; Hua, H.Y.; Jun, Z.N. A novel comprehensive evaluation methodology of organic Rankine cycle for parameters design and working fluid selection. *Appl. Therm. Eng.* **2018**, *143*, 283–292. [CrossRef]
28. Kolasiński, P. The Method of the Working Fluid Selection for Organic Rankine Cycle (ORC) Systems Employing Volumetric Expanders. *Energies* **2020**, *13*, 573. [CrossRef]

MDPI
St. Alban-Anlage 66
4052 Basel
Switzerland
Tel. +41 61 683 77 34
Fax +41 61 302 89 18
www.mdpi.com

Energies Editorial Office
E-mail: energies@mdpi.com
www.mdpi.com/journal/energies

www.ingramcontent.com/pod-product-compliance
Lightning Source LLC
Chambersburg PA
CBHW062050220926
51850CB00010B/2719
* 9 7 8 3 0 3 9 3 6 0 7 4 1 *